Parabolic Wave Equations with Applications

Michael D. Collins • William L. Siegmann

Parabolic Wave Equations with Applications

Springer

Michael D. Collins
Naval Research Laboratory
Washington, DC, USA

William L. Siegmann
Department of Mathematical Sciences
Rensselaer Polytechnic Institute
Troy, NY, USA

ISBN 978-1-4939-9932-3 ISBN 978-1-4939-9934-7 (eBook)
https://doi.org/10.1007/978-1-4939-9934-7

This Springer imprint is published by the registered company Springer Science+Business Media, LLC, part of Springer Nature.
The registered company address is: 233 Spring Street, New York, NY 10013, U.S.A.

To Jean and Nancy

Contents

About the Authors

Michael D. Collins was born in Greenville, Pennsylvania. He received his B.S. degree in mathematics from the Massachusetts Institute of Technology and his Ph.D. degree in applied mathematics from Northwestern University. Since 1985, he has worked for the Naval Research Laboratory. His research interests are ocean acoustics, wave propagation, inverse problems, and the ivory-billed woodpecker. Dr. Collins is a member of the Institute of Electrical and Electronics Engineers, the American Geophysical Union, and the Society for Industrial and Applied Mathematics.

William L. Siegmann was born in Pittsburgh, Pennsylvania. He received his B.S. and M.S. degrees in mathematics and his Ph.D. in applied mathematics from the Massachusetts Institute of Technology. From 1968 to 1970, he was a Postdoctoral Fellow in the Department of Mechanics at Johns Hopkins University. Since 1970, he has been in the Department of Mathematical Sciences at Rensselaer Polytechnic Institute. His research interests are ocean acoustics and wave propagation methods. Dr. Siegmann is a member of the IEEE Ocean Engineering Society, the Acoustical Society of America, and The Oceanography Society.

Chapter 1
Basic Concepts

Contents

1.1 Introduction

The parabolic equation method provides an appealing combination of accuracy and efficiency for many nonseparable wave propagation problems in geophysics. Parabolic wave equations are based on the assumption that range dependence (horizontal variations in the medium) is sufficiently gradual so that horizontally outgoing energy dominates energy that is back scattered toward the source. Appearing in Figs. 1.1 and 1.2 are some examples of range-dependent problems in ocean acoustics and seismology [1, 2]. Parabolic wave equations are derived from elliptic wave equations by expanding about a plane wave propagating in a preferred direction. The simplest expansions provide accurate solutions when the field is dominated by energy that propagates within a small angle of the preferred direction. For higher-order

M. D. Collins, W. L. Siegmann, *Parabolic Wave Equations with Applications*,
https://doi.org/10.1007/978-1-4939-9934-7_1

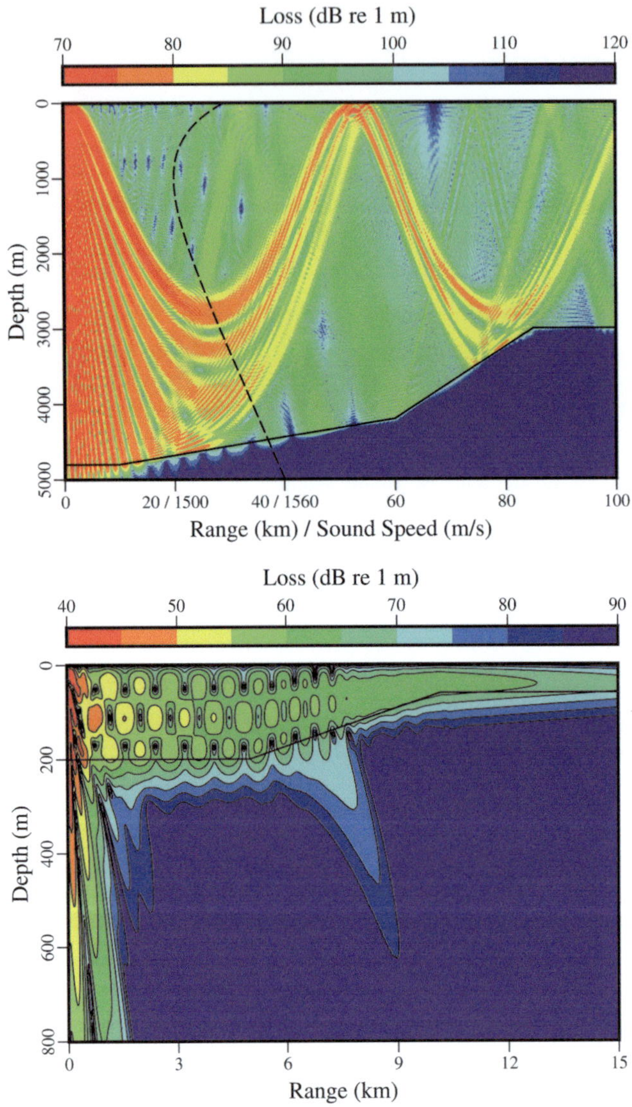

Fig. 1.1 Range-dependent problems in ocean acoustics. For the deep water problem (top), a 100 Hz source is 125 m below the surface, the canonical sound speed profile [1] has a minimum of 1500 m/s at a depth of 1000 m (dashed curve), and energy is refracted along curved paths and reflected from range-dependent bathymetry over the sediment, in which the sound speed is 1600 m/s and the density is 1.2 g/cm^3. For the shallow water problem (bottom), a 25 Hz source is 112 m below the surface, there is interference between modes, and one of the modes couples into a beam in the sediment [2], in which the sound speed is 1704.5 m/s and the density is 1.15 g/cm^3

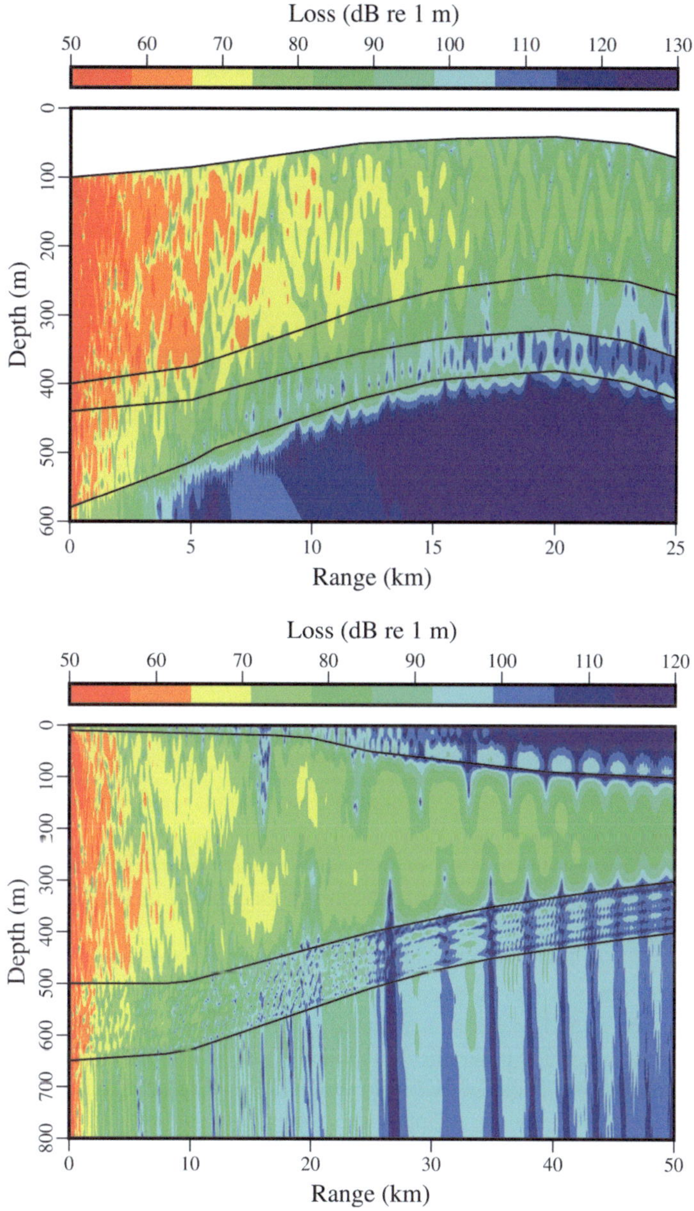

Fig. 1.2 Range-dependent problems in environments with elastic layers. In the sediment layers, the compressional speed ranges between 1500 and 3400 m/s, the shear speed ranges between 700 and 1700 m/s, and the density ranges between 1.5 and 2.5 g/cm^3. In the ice, the compressional speed is 3500 m/s, the shear speed is 1750 m/s, and the density is 0.9 g/cm^3. For the seismic problem (top), a 25 Hz source is 250 m below the surface, and the topography and layer thicknesses are range dependent. For the seismo-acoustics problem (bottom), a 25 Hz source is 140 m below an ice layer of variable thickness, which lies above an ocean and sediment layers

expansions, it is possible to handle wide propagation angles, and the only requirement is that outgoing energy dominates. Elliptic wave equations, such as the Helmholtz equation, correspond to boundary-value problems. Parabolic wave equations correspond to initial-value problems, which are much easier to solve. Solutions are marched outward in range by repeatedly solving systems of equations that involve banded matrices (tridiagonal for the acoustic case).

The concept of a parabolic wave equation is essential to some of the most basic problems in wave propagation. For example, the Green's function of the Helmholtz equation satisfies

$$\nabla^2 p + k^2 p = -4\pi\delta(\mathbf{x}), \tag{1.1}$$

where $\mathbf{x} = (x, y, z)$ is the position, k is the wave number, and $\delta(\mathbf{x})$ is the delta function. For the case of a homogeneous medium, the solution of Eq. (1.1) is the free space Green's function [3],

$$p = \frac{\exp(ikR)}{R}, \tag{1.2}$$

where $R = |\mathbf{x}|$. How can p be complex when k and $\delta(\mathbf{x})$ are real and the Helmholtz equation is linear? The answer is that Eq. (1.1) alone does not uniquely define p. The behavior of p for large R must also be specified. The free space Green's function is the solution that corresponds to the outgoing radiation condition,

$$\frac{\partial p}{\partial R} \cong ikp, \tag{1.3}$$

for $kR \gg 1$. This condition is a simple parabolic wave equation, with the preferred direction being away from the origin. Parabolic wave equations are often referred to simply as parabolic equations.

Leontovich and Fock pioneered the parabolic equation method in the 1940s and applied it to radio wave propagation in the atmosphere [4, 5]. Applications in seismology [6, 7] and ocean acoustics [8] in the 1970s marked the beginning of the widespread use of the parabolic equation method. This book covers progress that was made during the decades that followed. The remainder of this chapter covers background material on the elliptic wave equation and an introduction to the parabolic wave equation. The chapters that follow cover a variety of parabolic equation techniques and their applications.

1.2 The Elliptic Wave Equation

The parabolic equation method is applicable to range-dependent problems in which the parameters of the medium vary strongly with the depth z and gradually with the horizontal coordinates, but we begin by considering the range-independent case. In this section, we discuss basic techniques for solving the acoustic wave equation [9],

$$\nabla_{\perp}^2 p + \rho \frac{\partial}{\partial z}\left(\frac{1}{\rho}\frac{\partial p}{\partial z}\right) = \frac{1}{c^2}\frac{\partial^2 p}{\partial t^2}, \tag{1.4}$$

$$\nabla_{\perp}^2 = \nabla^2 - \frac{\partial^2}{\partial z^2}, \tag{1.5}$$

where p is the acoustic pressure, c is the sound speed, ρ is the density, and $\nabla_{\perp}^2$ is the horizontal component of the Laplacian. One of the applications of Eq. (1.4) is ocean acoustics problems involving a water column overlying a sediment, which may in some cases be modeled as an acoustic medium in which shear waves are neglected. The depth operator in Eq. (1.4) contains depth derivatives of the quantities p and $\rho^{-1}\partial p/\partial z$, which are continuous across interfaces, and it accounts for piecewise continuous depth variations in the environmental parameters ρ and c. The boundary condition $p = 0$ applies at the ocean surface $z = 0$ and deep within the sediment at $z = H$, where the domain is truncated. When ρ is constant, Eq. (1.4) reduces to the classical wave equation,

$$\nabla^2 p = \frac{1}{c^2}\frac{\partial^2 p}{\partial t^2}. \tag{1.6}$$

1.2.1 *The Frequency Domain*

In principle, Eq. (1.4) may be solved directly using numerical techniques that involve integrating forward in time. Since this approach tends to be impractical in large-scale environments, solutions are usually obtained by first taking advantage of aspects of the problem that permit simplification and then applying numerical techniques. Taking the Fourier transform of Eq. (1.4), we obtain the elliptic wave equation,

$$\nabla_{\perp}^2 \tilde{p} + \rho \frac{\partial}{\partial z}\left(\frac{1}{\rho}\frac{\partial \tilde{p}}{\partial z}\right) + k^2 \tilde{p} = 0, \tag{1.7}$$

$$\tilde{p}(\mathbf{x},\omega) = \int_{-\infty}^{\infty} p(\mathbf{x},t)\exp(i\omega t)dt, \tag{1.8}$$

where $k = \omega/c$ is the wave number. The problem has been reduced from four to three dimensions, but it is still very difficult in general. After solving Eq. (1.7), the Fourier transform may be inverted to obtain the time-domain solution,

$$p(\mathbf{x},t) = \frac{1}{2\pi}\int_{-\infty}^{\infty} \tilde{p}(\mathbf{x},\omega)\exp(-i\omega t)d\omega. \tag{1.9}$$

Since it will always be clear when we are working in the frequency domain, we will drop the tilde from $\tilde{p}$ to simplify the notation.

1.2.2 *The Uncoupled Azimuth Approximation*

For propagation from a point source in a waveguide, it is convenient to work with Eq. (1.7) in cylindrical geometry,

$$\frac{\partial^2 p}{\partial r^2} + \frac{1}{r}\frac{\partial p}{\partial r} + \frac{1}{r^2}\frac{\partial^2 p}{\partial \theta^2} + \rho\frac{\partial}{\partial z}\left(\frac{1}{\rho}\frac{\partial p}{\partial z}\right) + k^2 p = 0, \tag{1.10}$$

where r is the range and θ is the azimuth. The parabolic equation method is applicable when the environmental parameters vary gradually in the horizontal directions, which is the case for many problems in geophysics. When horizontal variations are sufficiently gradual, energy is approximately confined to planes of constant azimuth [10], and Eq. (1.10) simplifies to

$$\frac{\partial^2 p}{\partial r^2} + \frac{1}{r}\frac{\partial p}{\partial r} + \rho\frac{\partial}{\partial z}\left(\frac{1}{\rho}\frac{\partial p}{\partial z}\right) + k^2 p = 0. \tag{1.11}$$

The problem has been reduced to two dimensions, but it is still very difficult in general. Although Eq. (1.11) may be solved as a two-dimensional problem within a plane of constant θ, it provides three-dimensional solutions when k and ρ depend on θ. The uncoupled azimuth approximation makes it possible to routinely solve a wide range of problems in ocean acoustics and other areas with the parabolic equation method. Without this approximation, which is perhaps the most important in ocean acoustics, many problems of interest would require computations that would often be regarded as too costly and impractical to carry out.

1.2.3 The Far-Field Equation

Cylindrical spreading occurs when the properties of the medium vary sufficiently gradually in the horizontal directions so that amplitude behaves like $r^{-1/2}$. Substituting the new dependent variable $\tilde{p} = r^{1/2}p$ into Eq. (1.11), we obtain

$$\frac{\partial^2 \tilde{p}}{\partial r^2} + \rho \frac{\partial}{\partial z}\left(\frac{1}{\rho}\frac{\partial \tilde{p}}{\partial z}\right) + k^2\tilde{p} + \frac{1}{4r^2}\tilde{p} = 0. \tag{1.12}$$

For $kr \gg 1$, we neglect the $O(r^{-2})$ term in Eq. (1.12) and obtain the far-field equation,

$$\frac{\partial^2 p}{\partial r^2} + \rho \frac{\partial}{\partial z}\left(\frac{1}{\rho}\frac{\partial p}{\partial z}\right) + k^2 p = 0. \tag{1.13}$$

Since it will always be clear when we are using the far-field equation, we have dropped the tilde from $\tilde{p}$ to simplify the notation. For two-dimensional problems in plane geometry, Eq. (1.7) reduces to

$$\frac{\partial^2 p}{\partial x^2} + \rho \frac{\partial}{\partial z}\left(\frac{1}{\rho}\frac{\partial p}{\partial z}\right) + k^2 p = 0. \tag{1.14}$$

Since Eqs. (1.13) and (1.14) are similar, we will often consider two-dimensional problems in plane geometry for simplicity.

1.2.4 Source Conditions and Transmission Loss

In this section, we specify the normalization of source conditions for point and line sources. We include a point source term in Eq. (1.11) to obtain

$$\frac{\partial^2 p}{\partial r^2} + \frac{1}{r}\frac{\partial p}{\partial r} + \rho \frac{\partial}{\partial z}\left(\frac{1}{\rho}\frac{\partial p}{\partial z}\right) + k^2 p = -\frac{2}{r}\delta(r)\delta(z - z_0), \tag{1.15}$$

where z_0 is the source depth. The source terms in Eqs. (1.1) and (1.15) are normalized so that the free space solution has unit amplitude at a unit distance from the source. Including a line source term in Eq. (1.14), we obtain

$$\frac{\partial^2 p}{\partial x^2} + \rho \frac{\partial}{\partial z}\left(\frac{1}{\rho}\frac{\partial p}{\partial z}\right) + k^2 p = -2\tau\delta(x)\delta(z - z_0), \tag{1.16}$$

where $\tau = \sqrt{-2\pi i k(z_0)}$ is a normalization factor. For the case of a homogeneous medium and $z_0 = 0$, the solution of Eq. (1.16) is the free-space Green's function,

$$p(x,z) = \frac{i\tau}{2} H_0^{(1)}\left(k\left(x^2+z^2\right)^{1/2}\right). \tag{1.17}$$

From the choice of τ and the far-field approximation for the Hankel function, we obtain

$$p(x,z) \cong \left(x^2+z^2\right)^{-1/4} \exp\left(ik\left(x^2+z^2\right)^{1/2}\right). \tag{1.18}$$

Solutions of the wave equation are often displayed on a logarithmic scale in order to compress large variations in intensity. For sources normalized as in Eqs. (1.15) and (1.16), we define transmission loss in decibels to be $-20\log_{10}|p|$.

1.2.5 Lossy Media and Absorbing Layers

Attenuation is an important physical effect for many applications, and it is useful for truncating domains. For ocean acoustics problems, it is convenient to introduce artificial attenuation in an absorbing layer that is placed deep within the sediment to prevent spurious reflections from the boundary at $z = H$. Attenuation may be included in Eq. (1.7) by using the complex wave number,

$$\tilde{k} = k(1 + i\eta\sigma), \tag{1.19}$$

$$\eta = (40\pi \log_{10} e)^{-1}, \tag{1.20}$$

where σ is the attenuation in decibels per wavelength (dB/λ). To illustrate the purpose of the factor η, we consider a plane wave of the form,

$$p(x) = \exp\left(i\tilde{k}x\right), \tag{1.21}$$

and obtain

$$p(x + 2\pi/k) = \exp(-2\pi\eta\sigma)p(x), \tag{1.22}$$

where $2\pi/k$ is the wavelength. The increase in transmission loss over one wavelength is

$$-20\log_{10}(\exp(-2\pi\eta\sigma)) = 40\pi\eta\sigma\log_{10} e = \sigma. \tag{1.23}$$

In order to obtain numerical solutions of wave propagation problems, it is often useful to introduce an absorbing layer in order to prevent energy from reflecting from a boundary where the computational grid is truncated. An example in ocean acoustics is illustrated in Fig. 1.3. The depth of the ocean ranges between 60 and 200 m. The attenuation in the sediment is 0.5 dB per wavelength. When the grid is truncated

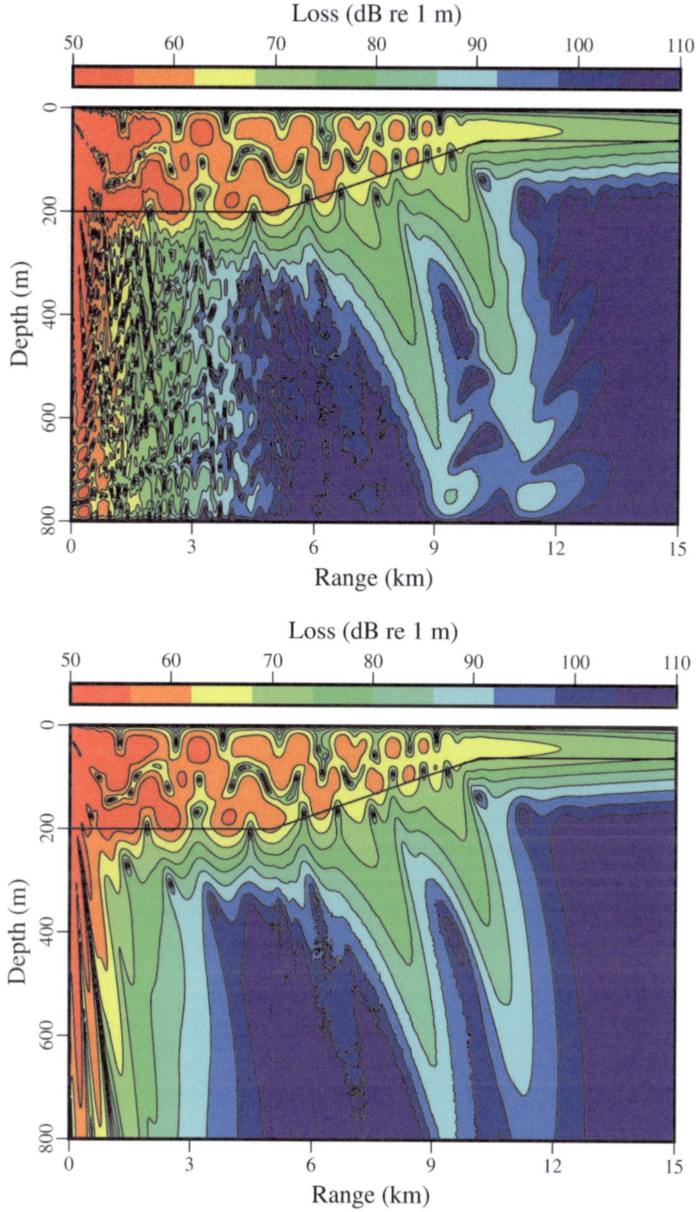

Fig. 1.3 Illustration of the use of an absorbing layer for the shallow water problem in Fig. 1.1, with the source depth modified to 180 m so that all three modes are excited. When the computational grid is truncated at a depth of 800 m (top), there are spurious reflections from the bottom boundary, some of which are visible in the water column for $r < 3$ km. When the domain extends to a depth of 2000 m and includes an absorbing layer (bottom), there are no spurious reflections

at $z = 800$ m, there are spurious reflections from the bottom boundary that affect the field in the water column for ranges within a few kilometers. When the grid is truncated at $z = 2000$ m and the attenuation is allowed to increase in the lower 500 m, there is no sign of spurious reflections.

1.2.6 The Normal Mode Solution

There exist two separation of variables solutions of Eq. (1.14). For the case in which k is real, the normal mode solution is of the form [3],

$$p(x,z) = \sum_m p_m(x)\zeta_m(z), \tag{1.24}$$

where $p_m(x)$ is proportional to $\exp(ik_m x)$ and the normal modes $\zeta_m(z)$ and eigenvalues k_m^2 satisfy

$$\rho \frac{d}{dz}\left(\frac{1}{\rho}\frac{d\zeta_m}{dz}\right) + k^2\zeta_m = k_m^2\zeta_m, \tag{1.25}$$

$$\int_0^H \rho^{-1}\zeta_m(z)\zeta_n(z)dz = \delta_{mn}. \tag{1.26}$$

Since the domain is finite and k is real, the spectrum consists of a discrete set of real eigenvalues of which a finite number are positive. The positive eigenvalues correspond to the propagating modes (they have small imaginary parts when attenuation is introduced). The negative eigenvalues correspond to the evanescent modes, which may be neglected for $kx \gg 1$. For a homogeneous medium, we obtain

$$\zeta_m(z) = \sqrt{\frac{2}{H}}\sin(kz\sin\phi_m), \tag{1.27}$$

$$k_m = k\cos\phi_m, \tag{1.28}$$

$$\sin\phi_m = \frac{m\pi}{kH}, \tag{1.29}$$

where ϕ_m is the propagation angle relative to the horizontal. For an inhomogeneous medium, the eigenvalues and modes may be obtained using numerical techniques. We illustrate the modes for a range-independent version of the shallow water problem in Fig. 1.1 in which the ocean depth is 200 m. The three propagating modes for this problem appear in Fig. 1.4. They are sinusoidal in the water column and exponentially decaying with depth in the sediment. The other case shown in Fig. 1.4 illustrates the effect that a discontinuity in density has on $\partial\zeta_m/\partial z$ at the interface.

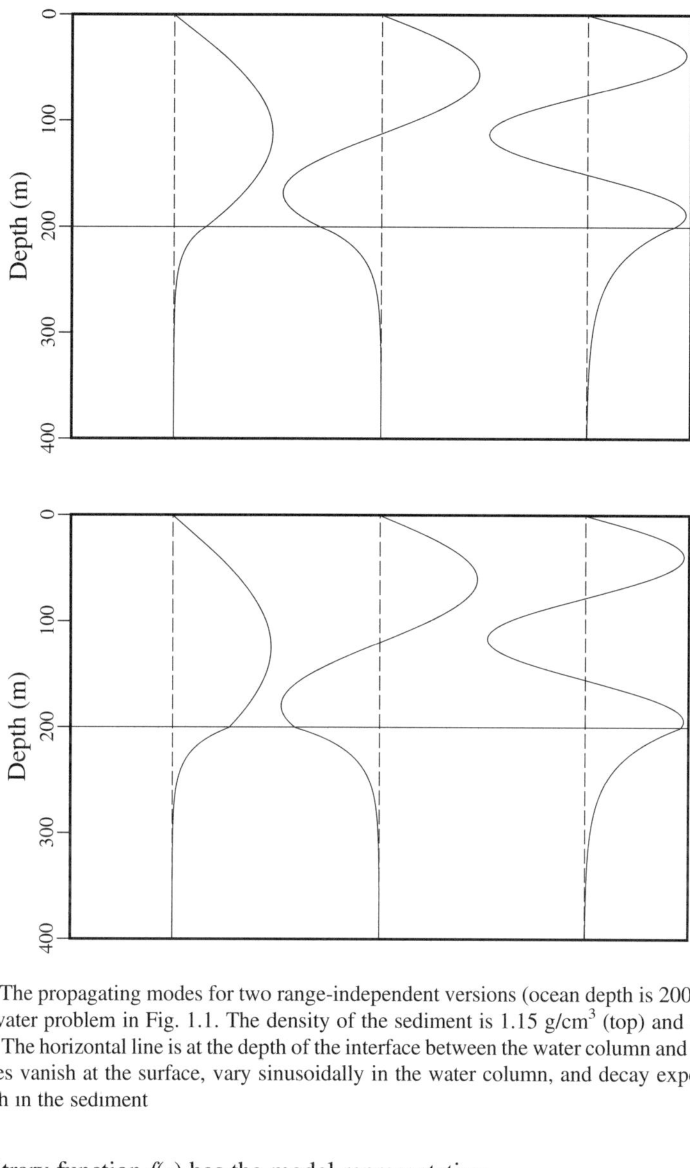

Fig. 1.4 The propagating modes for two range-independent versions (ocean depth is 200 m) of the shallow water problem in Fig. 1.1. The density of the sediment is 1.15 g/cm^3 (top) and 2.5 g/cm^3 (bottom). The horizontal line is at the depth of the interface between the water column and sediment. The modes vanish at the surface, vary sinusoidally in the water column, and decay exponentially with depth in the sediment

The arbitrary function $f(z)$ has the modal representation,

$$f(z) = \sum_m f_m \zeta_m(z), \tag{1.30}$$

$$f_m = \int_0^H \rho^{-1} f(z) \zeta_m(z) dz. \tag{1.31}$$

We assume unit density at the source depth and apply Eqs. (1.30) and (1.31) to obtain

$$\delta(z - z_0) = \sum_m \zeta_m(z_0)\zeta_m(z). \tag{1.32}$$

From Eqs. (1.16), (1.31), and (1.32), we obtain the line source solution,

$$\frac{d^2 p_m}{dx^2} + k_m^2 p_m = -2\tau\zeta_m(z_0)\delta(x), \tag{1.33}$$

$$p_m(x) = i\tau\zeta_m(z_0)k_m^{-1}\exp(ik_m|x|), \tag{1.34}$$

$$p(x, z) = i\tau\sum_m \zeta_m(z_0)\zeta_m(z)k_m^{-1}\exp(ik_m|x|). \tag{1.35}$$

Applying a similar analysis to Eq. (1.15), we obtain the point source solution,

$$\frac{d^2 p_m}{dr^2} + \frac{1}{r}\frac{dp_m}{dr} + k_m^2 p_m = -2\zeta_m(z_0)\frac{\delta(r)}{r}, \tag{1.36}$$

$$p_m(x) = i\pi\zeta_m(z_0)H_0^{(1)}(k_m r), \tag{1.37}$$

$$p(r, z) = i\pi\sum_m \zeta_m(z_0)\zeta_m(z)H_0^{(1)}(k_m r). \tag{1.38}$$

Applying the asymptotic expansion for the Hankel function, we obtain the far-field solution,

$$p(r, z) = \left(\frac{2\pi i}{r}\right)^{1/2}\sum_m \zeta_m(z_0)\zeta_m(z)k_m^{-1/2}\exp(ik_m r). \tag{1.39}$$

For ranges much greater than a wavelength, accurate solutions may be obtained by including only the propagating modes in Eqs. (1.35) and (1.39).

1.2.7 The Spectral Solution

The other separation of variables solution is based on integral transforms in range [3]. Taking the Fourier transform of Eq. (1.16), we obtain

$$\rho\frac{\partial}{\partial z}\left(\frac{1}{\rho}\frac{\partial\hat{p}}{\partial z}\right) + (k^2 - h^2)\hat{p} = -2\tau\delta(z - z_0), \tag{1.40}$$

$$\hat{p}(h,z) = \int_{-\infty}^{\infty} p(x,z)\exp(-ihx)dx, \tag{1.41}$$

where $\hat{p}$ is the wave-number spectrum and the separation constant h is the horizontal wave number. From Eqs. (1.25), (1.32), and (1.40), we obtain the modal representation of the wave-number spectrum,

$$\hat{p}(h,z) = 2\tau\sum_m \frac{\zeta_m(z_0)\zeta_m(z)}{h^2 - k_m^2}. \tag{1.42}$$

After solving Eq. (1.40), the Fourier transform may be inverted to obtain the spectral solution,

$$p(x,z) = \frac{1}{2\pi}\int_{-\infty}^{\infty} \hat{p}(h,z)\exp(ihx)dh. \tag{1.43}$$

When σ vanishes, the integration contour in Eq. (1.43) must be perturbed above the negative real axis and below the positive real axis in order to avoid the poles and obtain the physically meaningful solution [3]. Other than the finite-difference solution, the spectral solution is the most direct way of solving Eq. (1.16). The other separation of variables solution requires the solution of an eigenvalue problem. In cylindrical geometry, p and $\hat{p}$ are related by the Hankel transforms,

$$\hat{p}(h,z) = \int_0^{\infty} hJ_0(hr)p(r,z)dh, \tag{1.44}$$

$$p(r,z) = \int_0^{\infty} rJ_0(hr)\hat{p}(h,z)dr. \tag{1.45}$$

In the far field, the integral in Eq. (1.45) may be implemented using the asymptotic expression for the Bessel function. The wave number spectrum appears in Fig. 1.5 for the range-independent version of the shallow water problem in Fig. 1.1 in which the ocean depth is 200 m. The first and third modes are excited when the source is at $z = 112$ m. All three of the propagating modes are excited when the source is at $z = 180$ m.

1.3 The Parabolic Wave Equation

In this section, we derive parabolic wave equations and analyze their accuracy. We work with the elliptic wave equation in the form,

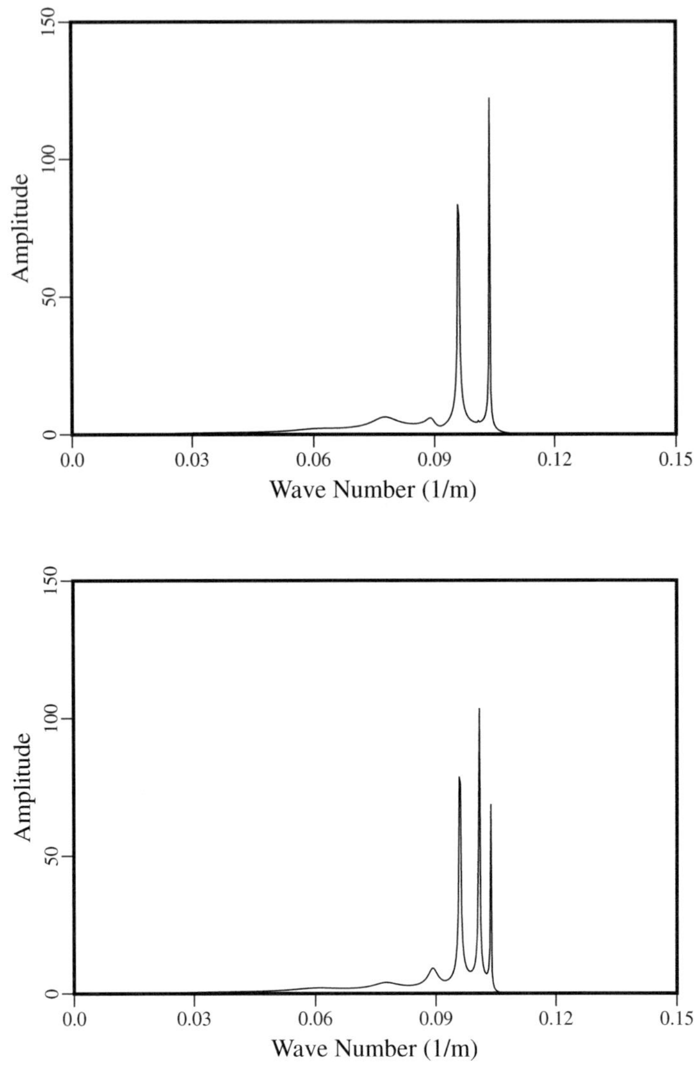

Fig. 1.5 The wave number spectrum at $z = 30$ m for a range-independent version (ocean depth is 200 m) of the shallow water problem in Fig. 1.1. The first and third modes are excited when the source depth is 112 m (top). All three of the propagating modes are excited when the source depth is 180 m (bottom)

$$\left(\frac{\partial^2}{\partial x^2} + k_0^2(1+X)\right)p = 0, \tag{1.46}$$

$$X = k_0^{-2}\left(\rho\frac{\partial}{\partial z}\frac{1}{\rho}\frac{\partial}{\partial z} + k^2 - k_0^2\right), \tag{1.47}$$

where k_0 is the reference wave number and $c_0 = \omega/k_0$ is the reference phase speed. Parabolic equations are based on the limit of X being small in the sense that $|Xp| \ll |p|$. This condition holds for the special case in which the medium is homogeneous and energy propagates nearly horizontally. It holds in general when $k_m \cong k_0$ for the propagating modes. The assumption of small X may be relaxed by using higher-order expansions in X.

1.3.1 Derivation by Factoring the Operator

Factoring the operator in Eq. (1.46), we obtain

$$\left(\frac{\partial}{\partial x} + ik_0(1+X)^{1/2}\right)\left(\frac{\partial}{\partial x} - ik_0(1+X)^{1/2}\right)p = 0. \tag{1.48}$$

Each factor in Eq. (1.48) corresponds to a parabolic equation. Solutions of the equation,

$$\left(\frac{\partial}{\partial x} - ik_0(1+X)^{1/2}\right)p = 0, \tag{1.49}$$

are also solutions of Eq. (1.48). Rearranging Eq. (1.49), we obtain the outgoing parabolic equation,

$$\frac{\partial p}{\partial x} = ik_0(1+X)^{1/2}p, \tag{1.50}$$

which corresponds to waves propagating in the positive x direction. The other factor in Eq. (1.48) gives rise to the incoming parabolic equation,

$$\frac{\partial p}{\partial x} = -ik_0(1+X)^{1/2}p, \tag{1.51}$$

which corresponds to waves propagating in the negative x direction.

Although solutions of Eq. (1.50) satisfy Eq. (1.46) exactly, it is necessary to introduce an approximation before the problem can be solved with numerical techniques. Applying a linear Taylor approximation to $(1 + X)^{1/2}$, we obtain the narrow-angle parabolic equation,

$$\frac{\partial p}{\partial x} = ik_0\left(1 + \frac{1}{2}X\right)p. \tag{1.52}$$

Although it is possible to achieve improved accuracy by including more terms of the Taylor series, rational approximations provide greater accuracy. Substituting a rational-linear approximation into Eq. (1.50), we obtain the wide-angle parabolic equation [7],

$$\frac{\partial p}{\partial x} = ik_0\left(1 + \frac{\frac{1}{2}X}{1 + \frac{1}{4}X}\right)p. \tag{1.53}$$

Substituting an n-term rational approximation into Eq. (1.50), we obtain the higher-order parabolic equation [11],

$$\frac{\partial p}{\partial x} = ik_0 R_n(X)p, \tag{1.54}$$

$$R_n(X) = 1 + \sum_{j=1}^{n} \frac{\alpha_{j,n}X}{1 + \beta_{j,n}X} = \prod_{j=1}^{n} \frac{1 + \gamma_{j,n}X}{1 + \beta_{j,n}X}. \tag{1.55}$$

Rational functions are used to approximate various functions that arise in parabolic equation techniques. The sum and product forms of $R_n(X)$ are both useful.

1.3.2 Derivation with Normal Modes

The normal mode solution is useful for precisely defining the operator square root. Since every solution of Eq. (1.46) can be represented in terms of the normal modes, a linear operator is defined once it is defined for each of the normal modes. From Eqs. (1.25) and (1.46), we obtain

$$k_0^2(1 + X)\zeta_m = k_m^2\zeta_m. \tag{1.56}$$

Repeated application of the operator $k_0^2(1 + X)$ to Eq. (1.56) leads to

$$\left(k_0^2(1 + X)\right)^N \zeta_m = \left(k_m^2\right)^N \zeta_m, \tag{1.57}$$

where N is an arbitrary integer. It follows from Eq. (1.57) and the Taylor series representation of the analytic function f that

$$f\left(k_0^2(1+X)\right)\zeta_m = f\left(k_m^2\right)\zeta_m. \tag{1.58}$$

As a special case of Eq. (1.58), we obtain

$$k_0(1+X)^{1/2}\zeta_m = k_m\zeta_m, \tag{1.59}$$

for $|X| < 1$. The physically meaningful analytic continuation of this definition corresponds to the branch of the square root function that maps the complex plane into the upper half plane so that evanescent waves decay with range. For some applications, parabolic equation solutions grow exponentially if the branch cut is not handled properly [12].

The parabolic equation can be derived using the normal mode representation. We consider a solution of Eq. (1.46) that consists of a single mode,

$$p(x,z) = \zeta_m(z)\exp\left(ik_m x\right). \tag{1.60}$$

Differentiating the exponential factor, we obtain

$$\frac{\partial p}{\partial x} = ik_m\zeta_m(z)\exp\left(ik_m x\right). \tag{1.61}$$

Applying Eq. (1.59), we obtain

$$\frac{\partial p}{\partial x} = ik_0(1+X)^{1/2}p. \tag{1.62}$$

Since every solution of Eq. (1.46) is a linear combination of modes, Eq. (1.62) holds for an arbitrary wave propagating in the positive x direction.

The normal mode solution is useful for analyzing the accuracy of parabolic equation solutions. Rearranging Eq. (1.56), we obtain

$$X\zeta_m = \frac{k_m^2 - k_0^2}{k_0^2}\zeta_m. \tag{1.63}$$

Repeating the argument used to derive Eq. (1.58), we obtain

$$f(X)\zeta_m = f\left(\frac{k_m^2 - k_0^2}{k_0^2}\right)\zeta_m. \tag{1.64}$$

Substituting a solution of the form,

$$p(x, z) = \zeta_m(z) \exp\left(i\tilde{k}_m x\right), \tag{1.65}$$

into Eq. (1.54) and using Eq. (1.64) for the case $f(X) = R_n(X)$, we obtain

$$\tilde{k}_m = k_0 R_n \left(\frac{k_m^2 - k_0^2}{k_0^2} \right). \tag{1.66}$$

We conclude that the solution of the parabolic equation is composed of the correct modes but not the correct eigenvalues. However, it follows from Eq. (1.66) that $\tilde{k}_m \cong k_m$ when

$$R_n(X) \cong (1 + X)^{1/2}. \tag{1.67}$$

1.4 Depth Numerics

The normal mode, spectral, and parabolic equation solutions require numerical solutions of equations of the form,

$$L\xi = f, \tag{1.68}$$

where the depth operator L may have depth-dependent coefficients, ξ is to be determined, and f is known. For the numerical approaches in this section, h is the grid spacing and the subscript j is used to designate numerical quantities at the jth grid point $z = z_j$.

1.4.1 Finite-Difference Formulas

The following finite-difference formulas, which were derived using Galerkin's method [13], are useful for approximating depth operators that arise in the applications:

$$\alpha\xi \cong \frac{\alpha_{j-1} + \alpha_j}{12}\xi_{j-1} + \frac{\alpha_{j-1} + 6\alpha_j + \alpha_{j+1}}{12}\xi_j + \frac{\alpha_j + \alpha_{j+1}}{12}\xi_{j+1}, \tag{1.69}$$

$$\frac{\partial \alpha}{\partial z}\xi \cong \frac{\alpha_j - \alpha_{j-1}}{6h}\xi_{j-1} + \frac{\alpha_{j+1} - \alpha_{j-1}}{3h}\xi_j + \frac{\alpha_{j+1} - \alpha_j}{6h}\xi_{j+1}, \tag{1.70}$$

$$\alpha\frac{\partial \xi}{\partial z} \cong -\frac{\alpha_{j-1} + 2\alpha_j}{6h}\xi_{j-1} + \frac{\alpha_{j-1} - \alpha_{j+1}}{6h}\xi_j + \frac{2\alpha_j + \alpha_{j+1}}{6h}\xi_{j+1}, \tag{1.71}$$

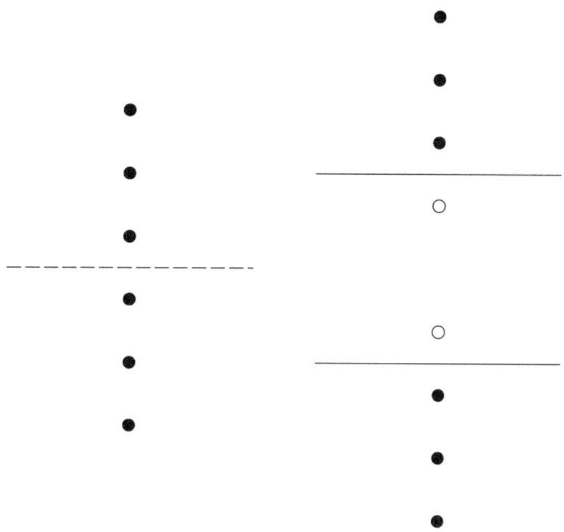

Fig. 1.6 A numerical approach for implementing source and interface conditions. The filled circles represent real grid points. The open circles represent artificial grid points. In the diagram on the left, the dashed horizontal line represents the depth of a source or an interface. In the diagram on the right, artificial grid points have been introduced just below the upper layer and just above the lower layer. Difference formulas for the source or interface conditions are used to determine the quantities at the artificial grid points in terms of the quantities at the real grid points

$$\frac{\partial}{\partial z}(\alpha\xi) \cong -\frac{2\alpha_{j-1}+\alpha_j}{6h}\xi_{j-1} + \frac{\alpha_{j+1}-\alpha_{j-1}}{6h}\xi_j + \frac{\alpha_j+2\alpha_{j+1}}{6h}\xi_{j+1}, \tag{1.72}$$

$$\frac{\partial \alpha}{\partial z}\frac{\partial \xi}{\partial z} \cong \frac{\alpha_{j-1}-\alpha_j}{2h^2}\xi_{j-1} - \frac{\alpha_{j-1}-2\alpha_j+\alpha_{j+1}}{2h^2}\xi_j + \frac{\alpha_{j+1}-\alpha_j}{2h^2}\xi_{j+1}, \tag{1.73}$$

$$\alpha\frac{\partial^2 \xi}{\partial z^2} \cong \frac{\alpha_j}{h^2}\xi_{j-1} - \frac{2\alpha_j}{h^2}\xi_j + \frac{\alpha_j}{h^2}\xi_{j+1}, \tag{1.74}$$

$$\frac{\partial}{\partial z}\left(\alpha\frac{\partial \xi}{\partial z}\right) \cong \frac{\alpha_{j-1}+\alpha_j}{2h^2}\xi_{j-1} - \frac{\alpha_{j-1}+2\alpha_j+\alpha_{j+1}}{2h^2}\xi_j + \frac{\alpha_j+\alpha_{j+1}}{2h^2}\xi_{j+1}, \tag{1.75}$$

where α is a coefficient function and each approximation has an $O(h^2)$ error. Finite-difference formulas for equations involving the operator X in Eq. (1.47) may be obtained by first dividing by ρ and then applying Eqs. (1.69) and (1.75).

1.4.2 *Source Conditions*

An approach that is illustrated in Fig. 1.6 may be used to handle the case in which $f(z) = \delta(z - z_0)$ in Eq. (1.68). Integrating Eq. (1.40) over an arbitrarily small interval about $z = z_0$ and assuming that ρ is constant near the source, we obtain

$$\frac{\partial \xi}{\partial z}\left(z_0^+\right) - \frac{\partial \xi}{\partial z}\left(z_0^-\right) = -2\tau. \tag{1.76}$$

The medium is divided into the layers $0 < z < z_0$ and $z_0 < z < H$, and the source depth is $z_0 = \frac{1}{2}(z_m + z_{m+1})$. We expand these layers slightly and introduce artificial grid points at $z = z_{m+1}$ in the upper layer and at $z = z_m$ in the lower layer. The artificial quantities $\tilde{\xi}_m$ and $\tilde{\xi}_{m+1}$ are assigned to the artificial grid points. At the grid points adjacent to the source, the finite-difference equations are of the form,

$$a_{1,m}\xi_{m-1} + a_{2,m}\xi_m + a_{3,m}\tilde{\xi}_{m+1} = 0, \tag{1.77}$$

$$a_{1,m+1}\tilde{\xi}_m + a_{2,m+1}\xi_{m+1} + a_{3,m+1}\xi_{m+2} = 0, \tag{1.78}$$

where $a_{i,j}$ are the entries of the tridiagonal matrix that approximates the depth operator. Applying approximations for continuity of ξ at $z = z_0$ and Eq. (1.76), we obtain

$$\frac{1}{2}\left(\xi_m + \tilde{\xi}_{m+1}\right) = \frac{1}{2}\left(\tilde{\xi}_m + \xi_{m+1}\right), \tag{1.79}$$

$$\frac{1}{h}\left(\xi_{m+1} - \tilde{\xi}_m\right) - \frac{1}{h}\left(\tilde{\xi}_{m+1} - \xi_m\right) = -2\tau. \tag{1.80}$$

Solving for the artificial values, we obtain

$$\tilde{\xi}_m = \xi_m + h\tau, \tag{1.81}$$

$$\tilde{\xi}_{m+1} = \xi_{m+1} + h\tau. \tag{1.82}$$

Substituting into Eqs. (1.77) and (1.78), we obtain

$$a_{1,m}\xi_{m-1} + a_{2,m}\xi_m + a_{3,m}\xi_{m+1} = -h\tau a_{3,m}, \tag{1.83}$$

$$a_{1,m+1}\xi_m + a_{2,m+1}\xi_{m+1} + a_{3,m+1}\xi_{m+2} = -h\tau a_{1,m+1}. \tag{1.84}$$

The source term gives rise to a term on the right side but does not affect the entries of the finite-difference approximation on the left side.

There is an alternative approach for implementing delta functions. When functions are approximated in terms of vectors in numerical solutions, the inner product of two such vectors (with a factor of h to account for dz) is a numerical approximation of the integral of the product of the functions. From the equation that expresses the basic property of the delta function,

$$\int \delta(z - z_0)\xi(z)dz = \xi(z_0), \tag{1.85}$$

we obtain

$$\delta(z - z_0) \cong \left(0, \ldots, 0, \frac{1}{h}, 0, \ldots, 0\right), \tag{1.86}$$

which contains a nonzero entry at the grid point $z = z_m \cong z_0$. Substituting this approximation into Eq. (1.85), we obtain

$$\int \delta(z - z_0)\xi(z)dz \cong \xi_m \cong \xi(z_0). \tag{1.87}$$

This approach is also applicable when the forcing function in Eq. (1.68) is a derivative of the delta function. For the case of the first derivative, we apply integration by parts to obtain

$$\int \delta'(z - z_0)\xi(z)dz = -\int \delta(z - z_0)\xi'(z)dz = -\xi'(z_0), \tag{1.88}$$

$$\delta'(z - z_0) \cong -\left(0, \ldots, 0, -\frac{1}{2h^2}, 0, \frac{1}{2h^2}, 0, \ldots, 0\right), \tag{1.89}$$

where the vector approximation contains nonzero entries at the grid points $z = z_{m-1}$ and $z = z_{m+1}$. Substituting into Eq. (1.88), we obtain

$$\int \delta'(z - z_0)\xi(z)dz \cong -\frac{\xi_{m+1} - \xi_{m-1}}{2h} \cong -\xi'(z_0). \tag{1.90}$$

For the case of the second derivative, we apply integration by parts twice and obtain

$$\int \delta''(z - z_0)\xi(z)dz = \int \delta(z - z_0)\xi''(z)dz = \xi''(z_0), \tag{1.91}$$

$$\delta''(z - z_0) \cong \left(0, \ldots, 0, \frac{1}{h^3}, -\frac{2}{h^3}, \frac{1}{h^3}, 0, \ldots, 0\right). \tag{1.92}$$

1.4.3 Interface Conditions

For the acoustic case, the Galerkin formulas are valid for problems involving piecewise continuous depth variations in the sound speed and density. For some of the other applications, discontinuities must be handled by explicitly applying interface conditions. The interface conditions may be implemented using an approach involving artificial grid points that is similar to the approach for implementing source conditions. We illustrate the approach for an acoustics problem involving an interface between layers A and B. The interface is located halfway between grid points m and $m+1$, the artificial quantity $\tilde{\xi}_{m+1}$ is defined in the upper layer, and the artificial quantity $\tilde{\xi}_m$ is defined in the lower layer. At the grid points adjacent to the interface, the finite-difference formulas for the depth operator are in the form,

$$a_{1,m}\xi_{m-1} + a_{2,m}\xi_m + a_{3,m}\tilde{\xi}_{m+1}, \tag{1.93}$$

$$a_{1,m+1}\tilde{\xi}_m + a_{2,m+1}\xi_{m+1} + a_{3,m+1}\xi_{m+2}. \tag{1.94}$$

To eliminate the artificial quantities, we apply the following finite-difference approximations for the interface conditions for continuity of p and $\rho^{-1}\partial p/\partial z$,

$$\frac{1}{2}\left(\xi_m + \tilde{\xi}_{m+1}\right) = \frac{1}{2}\left(\tilde{\xi}_m + \xi_{m+1}\right), \tag{1.95}$$

$$\frac{1}{\rho_A h}\left(\tilde{\xi}_{m+1} - \xi_m\right) = \frac{1}{\rho_B h}\left(\xi_{m+1} - \tilde{\xi}_m\right), \tag{1.96}$$

where the subscripts A and B denote evaluation in the respective layers. Solving for the artificial quantities, we obtain

$$\tilde{\xi}_m = b_{11}\xi_m + b_{12}\xi_{m+1}, \tag{1.97}$$

$$\tilde{\xi}_{m+1} = b_{21}\xi_m + b_{22}\xi_{m+1}, \tag{1.98}$$

where

$$b_{11} = \frac{2\rho_B}{\rho_A + \rho_B}, \tag{1.99}$$

$$b_{12} = \frac{\rho_A - \rho_B}{\rho_A + \rho_B}, \tag{1.100}$$

$$b_{21} = -\frac{\rho_A - \rho_B}{\rho_A + \rho_B}, \tag{1.101}$$

$$b_{22} = \frac{2\rho_A}{\rho_A + \rho_B}. \tag{1.102}$$

Using Eqs. (1.97) and (1.98) to eliminate the artificial quantities in Eqs. (1.93) and (1.94), we obtain

$$a_{1,m}\xi_{m-1} + \left(a_{2,m} + b_{21}a_{3,m}\right)\xi_m + b_{22}a_{3,m}\xi_{m+1}, \tag{1.103}$$

$$b_{11}a_{1,m+1}\xi_m + \left(a_{2,m+1} + b_{12}a_{1,m+1}\right)\xi_{m+1} + a_{3,m+1}\xi_{m+2}. \tag{1.104}$$

The interface may be handled by modifying entries of the tridiagonal matrix as follows:

$$a_{2,m} \rightarrow a_{2,m} + b_{21}a_{3,m}, \tag{1.105}$$

$$a_{3,m} \rightarrow b_{22}a_{3,m}, \tag{1.106}$$

$$a_{1,m+1} \rightarrow b_{11}a_{1,m+1}, \tag{1.107}$$

$$a_{2,m+1} \rightarrow a_{2,m+1} + b_{12}a_{1,m+1}. \tag{1.108}$$

1.4.4 Nonuniform Grid Spacing

The Galerkin formulas have been generalized to the case in which the grid spacing is nonuniform for the following cases [14]:

$$\begin{aligned}\gamma_j\alpha\xi \cong{}& \frac{h_j\left(\alpha_{j-1}+\alpha_j\right)}{12}\xi_{i-1} + \frac{h_j\left(\alpha_{j-1}+3\alpha_j\right)+h_{j+1}\left(3\alpha_j+\alpha_{j+1}\right)}{12}\xi_j \\ &+ \frac{h_{j+1}\left(\alpha_j+\alpha_{j+1}\right)}{12}\xi_{j+1},\end{aligned} \tag{1.109}$$

$$\begin{aligned}\gamma_j\frac{\partial}{\partial z}\left(\alpha\frac{\partial\xi}{\partial z}\right) \cong{}& \frac{\alpha_{j-1}+\alpha_j}{2h_j}\xi_{j-1} - \left(\frac{\alpha_{j-1}+\alpha_j}{2h_j}+\frac{\alpha_j+\alpha_{j+1}}{2h_{j+1}}\right)\xi_j \\ &+ \frac{\alpha_j+\alpha_{j+1}}{2h_{j+1}}\xi_{j+1},\end{aligned} \tag{1.110}$$

$$\gamma_j\alpha\frac{\partial\xi}{\partial z} \cong -\frac{\alpha_{j-1}+2\alpha_j}{6}\xi_{j-1} + \frac{\alpha_{j-1}-\alpha_{j+1}}{6}\xi_j + \frac{2\alpha_j+\alpha_{j+1}}{6}\xi_{j+1}, \tag{1.111}$$

$$\gamma_j\frac{\partial\alpha}{\partial z}\xi \cong \frac{\alpha_j-\alpha_{j-1}}{6}\xi_{j-1} + \frac{\alpha_{j+1}-\alpha_{j-1}}{3}\xi_j + \frac{\alpha_{j+1}-\alpha_j}{6}\xi_{j+1}, \tag{1.112}$$

where h_j and h_{j+1} are the grid spacings above and below z_j and $\gamma_j = \frac{1}{2}\left(h_j + h_{j+1}\right)$. Appearing in Fig. 1.7 are nearly identical solutions to the problem in Fig. 1.3 that were obtained using a uniform grid with 4000 points and an efficient nonuniform grid with a total of 500 points, with coarse sampling in the sediment and absorbing layer.

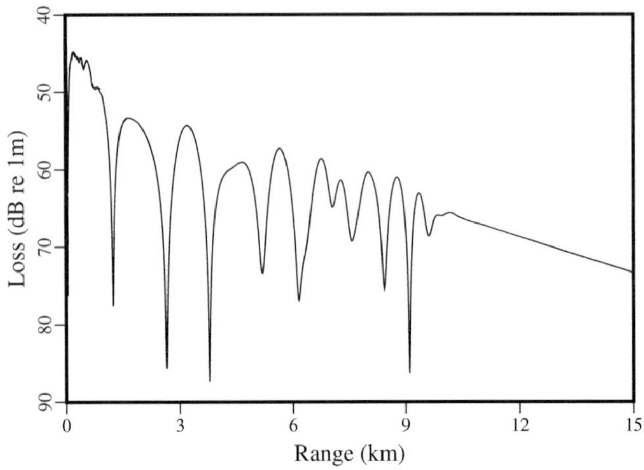

Fig. 1.7 Transmission loss at $z = 30$ m for the shallow water problem in Fig. 1.3. The dashed curve was obtained using 4000 grid points at a uniform spacing of 0.5 m. The nearly identical solid curve was obtained using 500 grid points, with 420 spaced at 0.5 m in the water column and upper sediment, 40 spaced at 5 m starting 10 m into the sediment, and 40 spaced at 40 m in the lower sediment and absorbing layer

References

1. W.H. Munk, "Sound channel in an exponentially stratified ocean, with applications to SOFAR," J. Acoust. Soc. Am. **55**, 220–226 (1974).
2. F.B. Jensen and W.A. Kuperman, "Sound propagation in a wedge-shaped ocean with a penetrable bottom," J. Acoust. Soc. Am. **67**, 1564–1566 (1980).
3. F.B. Jensen, W.A. Kuperman, M.B. Porter, and H. Schmidt, *Computational Ocean Acoustics* (American Institute of Physics, New York, 1994).
4. M.A. Leontovich and V.A. Fock, "Solution of the problem of propagation of electromagnetic waves along the earth's surface by the method of parabolic equation," J. Exp. Theor. Phys. **16**, 557–573 (1946).
5. V.A. Fock, *Electromagnetic Diffraction and Propagation Problems*, Pergamon, New York (1965).
6. J.F. Claerbout, "Coarse grid calculations of waves in inhomogeneous media with application to delineation of complicated seismic structure," Geophysics **35**, 407–418 (1970).
7. J.F. Claerbout, *Fundamentals of Geophysical Data Processing*, McGraw-Hill, New York (1976).
8. F.D. Tappert, "The parabolic approximation method," in *Wave Propagation and Underwater Acoustics*, edited by J.B. Keller and J.S. Papadakis (Springer, New York, 1977).
9. P.G. Bergmann, "The wave equation in a medium with a variable index of refraction," J. Acoust. Soc. Am. **17**, 329–333 (1946).
10. J.S. Perkins and R.N. Baer, "An approximation to the three-dimensional parabolic-equation method for acoustic propagation," J. Acoust. Soc. Am. **72**, 515–522 (1982).
11. A. Bamberger, B. Engquist, L. Halpern, and P. Joly, "Higher order paraxial wave equation approximations in heterogeneous media," SIAM J. Appl. Math. **48**, 129–154 (1988).
12. B.T.R. Wetton and G.H. Brooke, "One-way wave equations for seismoacoustic propagation in elastic waveguides," J. Acoust. Soc Am. **87**, 624–632 (1990).
13. M.D. Collins, "A higher-order parabolic equation for wave propagation in an ocean overlying an elastic bottom," J. Acoust. Soc. Am. **86**, 1459–1464 (1989).
14. W.M. Sanders and M.D. Collins, "Nonuniform depth grids in parabolic equation solutions," J. Acoust. Soc. Am. **133**, 1953–1958 (2013).

Chapter 2
Parabolic Equation Techniques

Contents

2.1 Introduction

Parabolic equation techniques are based on rational approximations of the square root and other functions. The split-step Padé solution is based on an exponential of the square root [1], which takes into account the range numerics and allows large range steps. An initial condition may be obtained using the self-starter [2], which is based on a far-field approximation of a Hankel function of the square root. Accurate solutions to range-dependent problems may be obtained with the energy-conserving parabolic equation [3], which is based on the fourth root. Three-dimensional parabolic equations are derived by factoring the wave equation without making the uncoupled azimuth approximation [4–6]. When horizontal variations in the environment are sufficiently gradual so that energy coupling between modes may be neglected, three-dimensional calculations can be avoided by solving horizontal wave equations for the mode coefficients [7, 8].

M. D. Collins, W. L. Siegmann, *Parabolic Wave Equations with Applications*,
https://doi.org/10.1007/978-1-4939-9934-7_2

2.2 Rational Approximations

There are various choices for the coefficients of the rational approximation $R_n(X)$ in Eq. (1.55). A Padé approximation of the function $f(X)$ corresponds to matching the first $2n$ derivatives at $X = 0$. For the case $f(X) = (1 + X)^{1/2}$, analytic expressions are known for the Padé coefficients [9, 10],

$$\alpha_{j,n} = \frac{2}{2n+1} \sin^2\left(\frac{j\pi}{2n+1}\right), \tag{2.1}$$

$$\beta_{j,n} = \cos^2\left(\frac{j\pi}{2n+1}\right), \tag{2.2}$$

$$\gamma_{j,n} = \sin^2\left(\frac{j\pi}{2n+1}\right). \tag{2.3}$$

Since Padé approximations provide a high level of accuracy near $X = 0$, the propagating modes can often be handled accurately by using just a few terms. For the case $k(z) = k_0$, we substitute Eq. (1.28) into Eq. (1.66) and obtain

$$\tilde{k}_m = k_0 R_n\left(-\sin^2\phi_m\right), \tag{2.4}$$

$$\varepsilon_m = \left|\frac{k_m - \tilde{k}_m}{k_m}\right| = \left|1 - \frac{R_n\left(-\sin^2\phi_m\right)}{\cos\phi_m}\right|, \tag{2.5}$$

where ε_m is the relative error.

2.2.1 *Numerically Generated Rational Approximations*

Since the Padé coefficients are real, it follows from Eq. (1.66) that all modes, including the evanescent modes, are treated as propagating modes, and this is a cause of stability issues for some applications. Numerical techniques can be used to derive rational approximations that accurately handle the propagating modes and annihilate the evanescent modes. Determining the coefficients is a nonlinear problem that can be solved efficiently by first solving a linear problem and then finding the roots of a polynomial [11]. The linear problem is based on the representation,

$$f(X) \cong R_n(X) = f(0)\frac{1 + \sum_{j=1}^{n} \tilde{\gamma}_{j,n} X^j}{1 + \sum_{j=1}^{n} \tilde{\beta}_{j,n} X^j}. \tag{2.6}$$

Effective choices for the coefficients may be defined in terms of accuracy and stability constraints [12]. The purpose of the accuracy constraints is to ensure that the rational approximation is accurate near $X = 0$. An effective choice for these constraints is to require derivatives of the error function,

$$E(X) = f(X)\left(1 + \sum_{j=1}^{n} \tilde{\beta}_{j,n} X^j\right) - f(0)\left(1 + \sum_{j=1}^{n} \tilde{\gamma}_{j,n} X^j\right), \tag{2.7}$$

to vanish at $X = 0$. The purpose of the stability constraints is to ensure that the evanescent eigenvalues are mapped into the upper half plane so that the evanescent modes decay with range. These constraints are applied in the region to the left of the branch point $X = -1$, where the evanescent eigenvalues are located. After the linear problem for the coefficients $\tilde{\beta}_{j,n}$ and $\tilde{\gamma}_{j,n}$ is solved (by least squares if there are more than $2n$ constraints), the coefficients $\beta_{j,n}$ and $\gamma_{j,n}$ in Eq. (1.55) are obtained by finding the roots of the polynomials in Eq. (2.6). Effective algorithms exist for the nonlinear problem of finding roots [13], which is easier to solve than the nonlinear problem of directly searching for $\beta_{j,n}$ and $\gamma_{j,n}$.

Various rational approximations are compared in Fig. 2.1 for the case $f(X) = (1 + X)^{1/2}$. The linear Taylor approximation starts to break down near $\phi_m = 15^\circ$. The one-term Padé approximation starts to break down near $\phi_m = 30^\circ$. For the numerically generated rational approximations, the accuracy constraints are that the first $2n - 1$ derivatives of $E(X)$ vanish at $X = 0$, and the stability constraint $R_n(-3) = i$ forces the approximation to map $X < -1$ into the upper half of the complex plane so that evanescent modes are annihilated. Including one stability constraint results in a significant loss of accuracy for the case $n = 1$ but has a smaller effect for the other cases. The image of the segment $-5 < X < 3$ of the real line appears in Fig. 2.2 for the cases $n = 4$ and 8. Although the evanescent modes do not decay at the correct rates under these mappings, all that matters for some applications is that they decay rapidly enough to avoid stability problems.

2.2.2 *Rotated Rational Approximations*

For the case $f(X) = (1 + X)^{1/2}$, the Padé approximation can be modified to annihilate the evanescent modes by rotating the branch cut $X < -1$ about $X = -1$ [14]. To rotate the branch cut by the angle ν, we define $\tilde{X}$ by

$$1 + X = e^{i\nu}\left(1 + \tilde{X}\right), \tag{2.8}$$

and obtain

$$(1 + X)^{1/2} = e^{i\nu/2}\left(1 + \tilde{X}\right)^{1/2}. \tag{2.9}$$

Approximating the square root on the right side of Eq. (2.9) by a rational function in $\tilde{X}$, we obtain

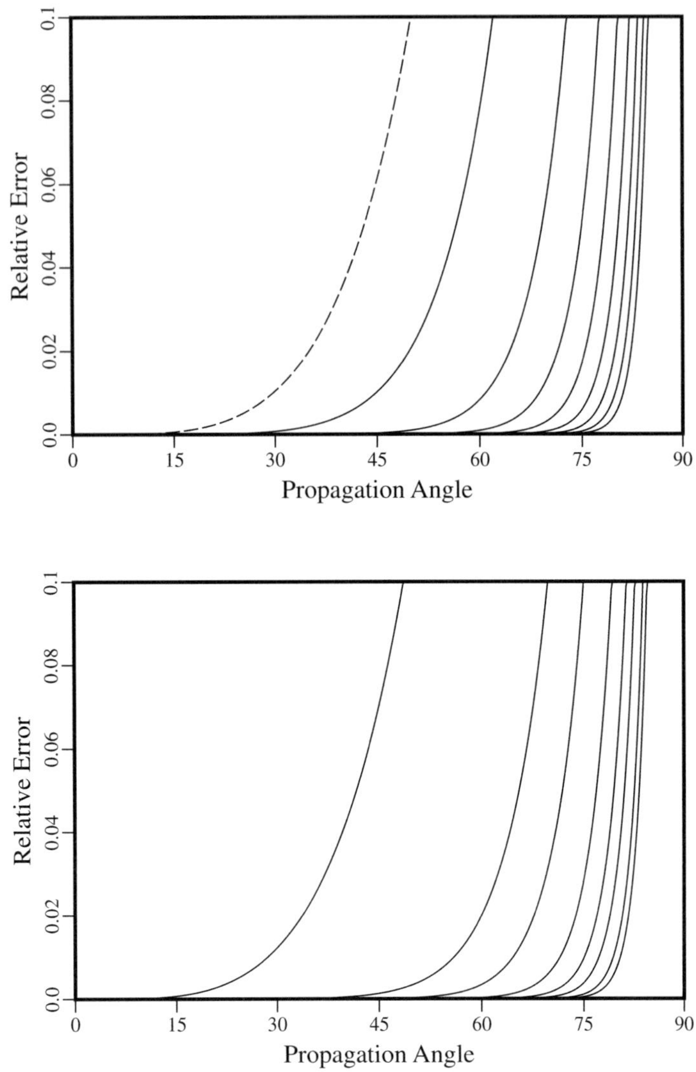

Fig. 2.1 Comparison of rational approximations of the square root function. The upper graph shows the linear Taylor approximation (dashed curve) and Padé approximations for $n \leq 8$. The lower graph shows rational approximations for $n \leq 8$ that were generated using $2n - 1$ accuracy constraints and one stability constraint

$$(1+X)^{1/2} = e^{i\nu/2}\left(1+\tilde{X}\right)^{1/2} \cong e^{i\nu/2}\prod_{j=1}^{n}\frac{1+\gamma_{j,n}\tilde{X}}{1+\beta_{j,n}\tilde{X}}, \tag{2.10}$$

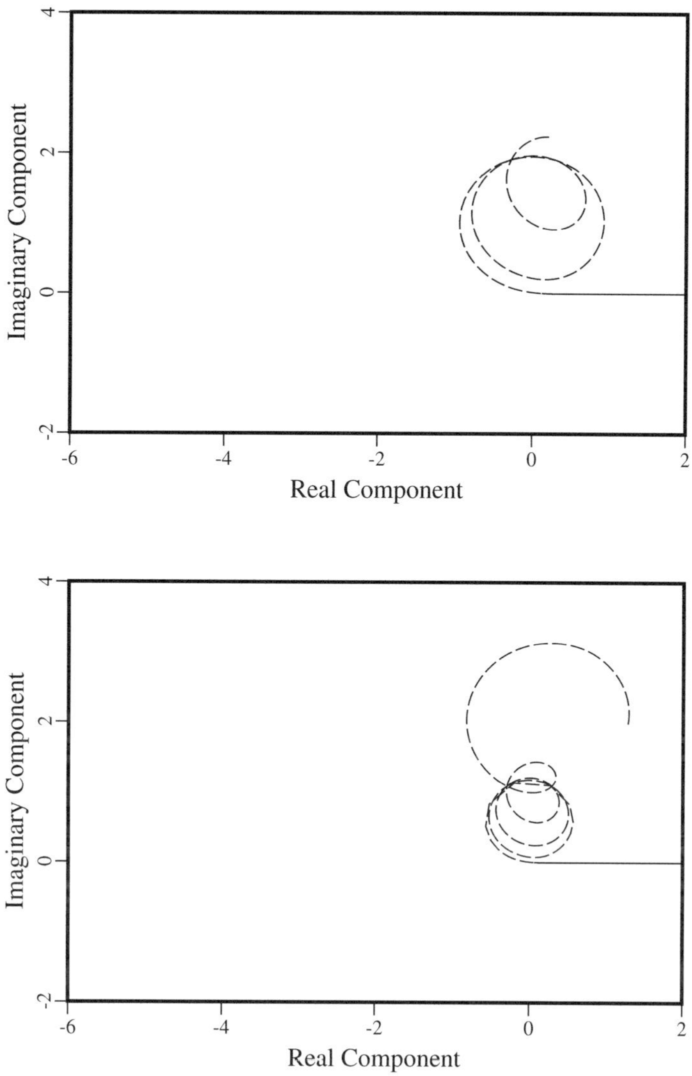

Fig. 2.2 Images of part of the real line under numerically generated rational approximations for $n = 4$ (top) and $n = 8$ (bottom). The dashed parts of the curves are the images of $-5 < X < -1$. The solid parts of the curves are the images of $-1 < X < 3$

where $\beta_{j,n}$ and $\gamma_{j,n}$ may be the coefficients in Eqs. (2.2) and (2.3) or any other rational approximation of the square root function. Using Eq. (2.8) to eliminate $\tilde{X}$ in Eq. (2.10), we obtain

$$(1+X)^{1/2} \cong e^{i\nu/2} \prod_{j=1}^{n} \frac{1+\gamma_{j,n}(e^{-i\nu}-1)}{1+\beta_{j,n}(e^{-i\nu}-1)} \prod_{j=1}^{n} \frac{1+\tilde{\gamma}_{j,n}X}{1+\tilde{\beta}_{j,n}X}, \tag{2.11}$$

$$\tilde{\beta}_{j,n} = \frac{e^{-i\nu}\beta_{j,n}}{1+\beta_{j,n}(e^{-i\nu}-1)}, \tag{2.12}$$

$$\tilde{\gamma}_{j,n} = \frac{e^{-i\nu}\gamma_{j,n}}{1+\gamma_{j,n}(e^{-i\nu}-1)}. \tag{2.13}$$

It follows from Eq. (2.10) that the first product in Eq. (2.11) is approximately $e^{-i\nu/2}$, and we obtain the rotated rational approximation,

$$(1+X)^{1/2} \cong \prod_{j=1}^{n} \frac{1+\tilde{\gamma}_{j,n}X}{1+\tilde{\beta}_{j,n}X}. \tag{2.14}$$

Applying the same approach for the sum form of the rational approximation, we obtain

$$(1+X)^{1/2} \cong e^{i\nu/2}\left[1+\sum_{j=1}^{n} \frac{\alpha_{j,n}(e^{-i\nu}-1)}{1+\beta_{j,n}(e^{-i\nu}-1)}\right] + \sum_{j=1}^{n} \frac{\tilde{\alpha}_{j,n}X}{1+\tilde{\beta}_{j,n}X}, \tag{2.15}$$

$$\tilde{\alpha}_{j,n} = \frac{e^{-i\nu/2}\alpha_{j,n}}{\left(1+\beta_{j,n}(e^{-i\nu}-1)\right)^2}. \tag{2.16}$$

The term in brackets in Eq. (2.15) is approximately $e^{-i\nu/2}$, and we obtain the rotated rational approximation,

$$(1+X)^{1/2} \cong 1+\sum_{j=1}^{n} \frac{\tilde{\alpha}_{j,n}X}{1+\tilde{\beta}_{j,n}X}. \tag{2.17}$$

The rotated rational approximations can be expressed in analytic form when applied to the coefficients in Eqs. (2.1)–(2.3), and they become more effective at annihilating evanescent modes as the rotation angle increases. Accuracy decreases as the rotation angle increases, however, since the location of the accuracy constraints moves from $X = 0$ to $\tilde{X} = 0$, which corresponds to $X = e^{i\nu} - 1$. Two of the rotated mappings appear in Fig. 2.3. For the case $n = 4$ and $\nu = 20^{\circ}$, the behavior is similar to the numerically generated rational approximations in Fig. 2.2. For the case $n = 8$ and $\nu = 90^{\circ}$, most of the segment $-5 < X < 3$ is properly mapped (the exception being a small neighborhood of the branch point $X = -1$). The relative error appears in Fig. 2.4 for two cases. For $\nu = 20^{\circ}$, accuracy is comparable to the numerically generated rational approximations that appear in Fig. 2.1. For $\nu = 90^{\circ}$, there is a

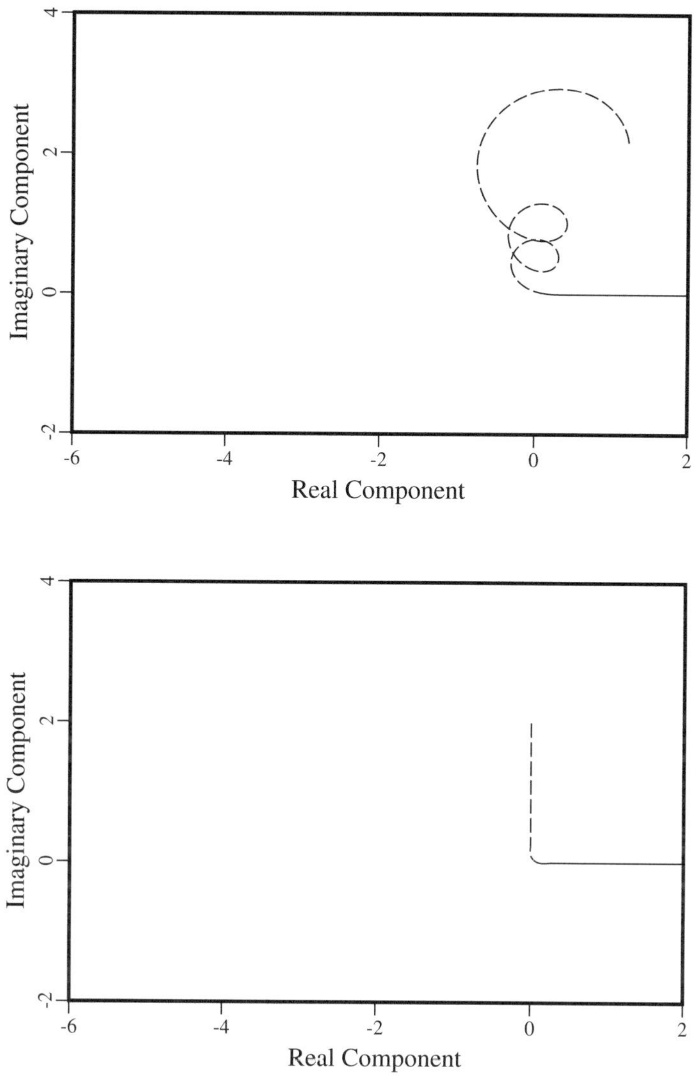

Fig. 2.3 Images of part of the real line under rotated rational approximations for $n = 4$ and $\nu = 20^{\circ}$ (top) and $n = 8$ and $\nu = 90^{\circ}$ (bottom). The dashed parts of the curves are the images of $-5 < X < -1$. The solid parts of the curves are the images of $-1 < X < 3$

significant loss of accuracy for $n \leq 4$, but these approximations are very effective for larger n.

The rotated rational approximations often provide improved stability, but in some cases they can actually introduce stability problems. As we illustrate for a problem involving a 500 m thick waveguide in which $c = 1500$ m/s, there are cases in which the rational approximations of Ref. [9] provide stable solutions but the rotated

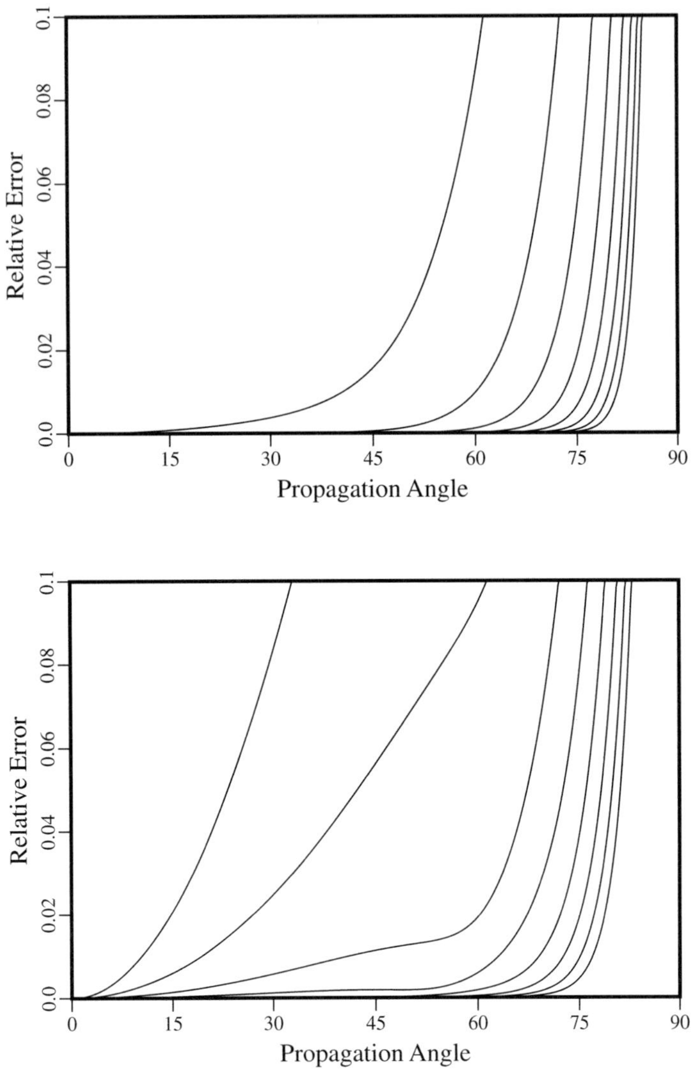

Fig. 2.4 Comparison of rotated rational approximations of the square root function. The upper graph shows the case $\nu = 20^{\circ}$ for $n \leq 8$. The lower graph shows the case $\nu = 90^{\circ}$ for $n \leq 8$

rational approximations provide unstable solutions. As shown in Fig. 2.5, a small interval of the real line (near the branch point) is mapped into the lower half plane for the case $n = 10$ and $\nu = 110^{\circ}$. At 50 Hz, all of the eigenvalues are mapped on or above the real line. The eigenvalues are more densely spaced at 250 Hz, and two of them are mapped below the real line. Stability problems associated with these eigenvalues, which correspond to energy that propagates nearly vertically, can be avoided by introducing attenuation and/or using a smaller rotation angle.

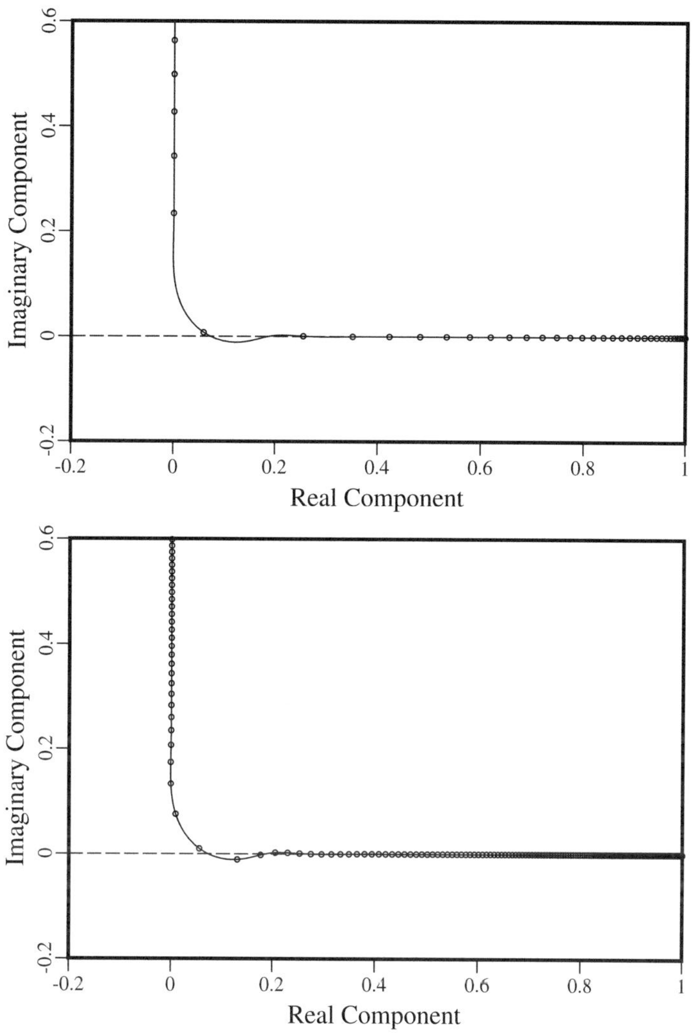

Fig. 2.5 Mappings of the eigenvalues (small circles) of problems involving a homogeneous waveguide under the rotated rational approximation for $n = 10$ and $\nu = 110^\circ$. In a small interval near the branch point, the image of the real line (solid curve) is in the lower half plane. For the 50 Hz case (top), all of the eigenvalues are mapped on or above the real line. The eigenvalues are more densely spaced for the 250 Hz case (bottom), and two of them are mapped below the real line

2.3 The Split-Step Finite-Difference Solution

The solution of Eq. (1.54) may be obtained with the splitting method [15], which requires numerical integrations over the range step Δx of the equations,

$$\frac{\partial p}{\partial x} = ik_0 p, \tag{2.18}$$

$$\frac{\partial p}{\partial x} = \frac{ik_0 \alpha_{j,n} X}{1 + \beta_{j,n} X} p. \tag{2.19}$$

Integrating Eq. (2.18) exactly and applying Crank-Nicolson integration to Eq. (2.19), we obtain the split-step finite-difference solution,

$$p(x + \Delta x) = \exp(ik_0 \Delta x) \prod_{j=1}^{n} \frac{1 + \tilde{\gamma}_{j,n} X}{1 + \tilde{\beta}_{j,n} X} p(x), \tag{2.20}$$

$$\tilde{\beta}_{j,n} = \beta_{j,n} - \frac{1}{2} ik_0 \Delta x \alpha_{j,n}, \tag{2.21}$$

$$\tilde{\gamma}_{j,n} = \beta_{j,n} + \frac{1}{2} ik_0 \Delta x \alpha_{j,n}. \tag{2.22}$$

Expanding the rational function in Eq. (2.20) by the method of partial fractions, we obtain an equation of the form,

$$p(x + \Delta x) = \exp(ik_0 \Delta x) \left(1 + \sum_{j=1}^{n} \frac{\tilde{\alpha}_{j,n} X}{1 + \tilde{\beta}_{j,n} X} \right) p(x), \tag{2.23}$$

which may be solved efficiently on a parallel-processing computer by assigning the jth term to the jth processor. Solutions generated with the split-step finite-difference solution and Padé approximations are compared with the normal mode solution in Fig. 2.6 for an example involving a 25 Hz source at $z = 25$ m in a waveguide in which $c = 1500$ m/s, with the boundary condition $p = 0$ at $z = 0$ and $z = 250$ m. The modes propagate at angles of approximately 7°, 14°, 21°, 29°, 37°, 46°, 57°, and 74° [10]. Errors start to appear in the $n = 4$ solution for $r > 1$ km. The $n = 6$ solution is nearly identical to the normal mode solution.

The solution is advanced in range by repeatedly solving tridiagonal systems of equations. The parabolic equation can be applied to range-dependent problems by updating the tridiagonal matrices as the acoustic parameters vary with range. The normal mode and spectral solutions can also be applied to range-dependent problems by updating the modes [16] and spectrum [17], but the parabolic equation method is more efficient than these approaches since the solution is not decomposed into the modes or spectrum. The efficiency of this approach can be improved by storing the LU decomposition of the matrices [18] in order to avoid repeating the same

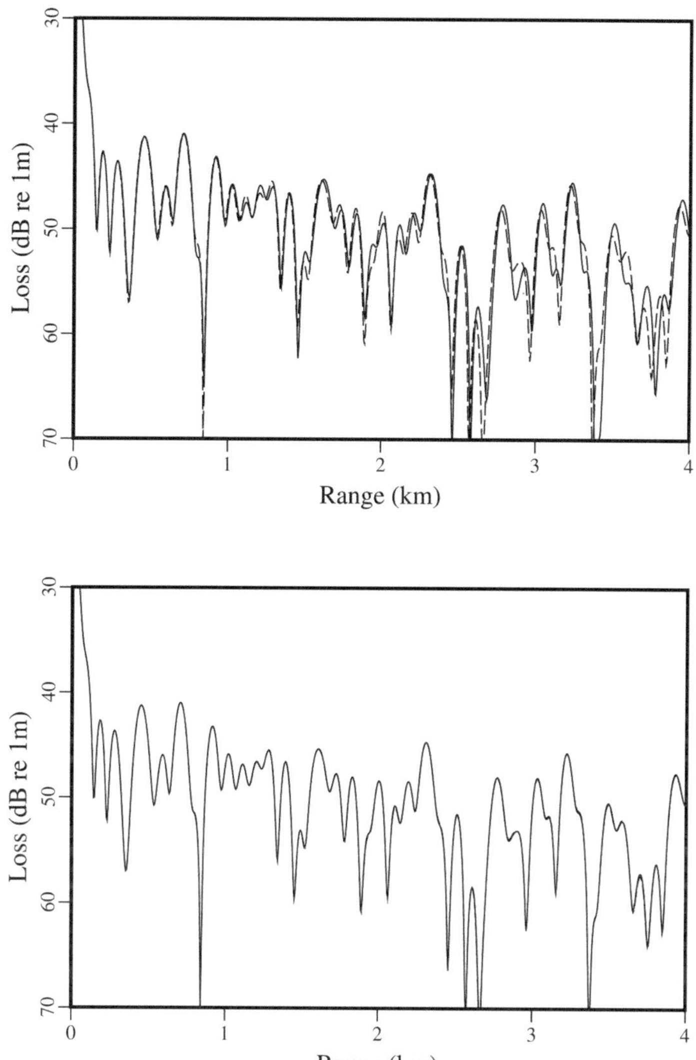

Fig. 2.6 Transmission loss at $z = 25$ m for a problem involving a homogeneous waveguide with perfectly reflecting boundaries and modes with propagation angles of 7°, 14°, 21°, 29°, 37°, 46°, 57°, and 74°. The solid curves were obtained with the split-step finite-difference solution for the cases $n = 4$ (top) and $n = 6$ (bottom). The dashed curves are the normal mode solution

calculations at each range step. Gaussian elimination is not the most efficient approach for problems with range dependence of the form,

$$k(x,z) = \begin{cases} k_1(z) & \text{for } z < d(x) \\ k_2(z) & \text{for } z > d(x), \end{cases} \tag{2.24}$$

where the bathymetry $d(x)$ is the depth of a sloping interface. The first step of this algorithm is to eliminate entries below the main diagonal by sweeping downward. In this process, the entries of the matrix and the vector are both modified. For a range-independent problem, this may be done efficiently by saving the decomposition of the matrix that takes place during the elimination step. If the bathymetry changes, however, the decomposition must be repeated throughout the entire sediment $z > d(x)$. The approach illustrated in Fig. 2.7 is more efficient than Gaussian elimination for this problem [19]. There are two elimination steps, which involve sweeping toward the interface from above and below. Entries below the main diagonal are eliminated on the downward sweep. Entries above the main diagonal are eliminated on the upward sweep. When the bathymetry changes, it is necessary to update only a few entries of the decomposed matrix. The efficiency of this algorithm is not affected by variations in bathymetry, and it is applicable to banded systems involving more than three diagonals.

2.4 The Split-Step Padé Solution

The split-step finite-difference solution is orders of magnitude more efficient than the finite-difference solution of the elliptic wave equation. The split-step Padé solution provides an additional gain in efficiency by incorporating the range numerics into the rational approximation. We assume that the acoustic parameters are range-independent over the interval $(x, x+\Delta x)$. From Eq. (1.24), we obtain

$$p(x+\Delta x, z) = \sum_m p_m(x)\exp(ik_m\Delta x)\zeta_m(z). \tag{2.25}$$

From Eqs. (1.58) and (2.25), we obtain

$$p(x+\Delta x, z) = \exp\left(ik_0\Delta x(1+X)^{1/2}\right)\sum_m p_m(x)\zeta_m(z), \tag{2.26}$$

$$p(x+\Delta x) = \exp\left(ik_0\Delta x(1+X)^{1/2}\right)p(x). \tag{2.27}$$

Substituting a rational approximation for the exponential function in Eq. (2.27), we obtain the split-step Padé solution [1],

$$p(x+\Delta x) = R_n(X)p(x), \tag{2.28}$$

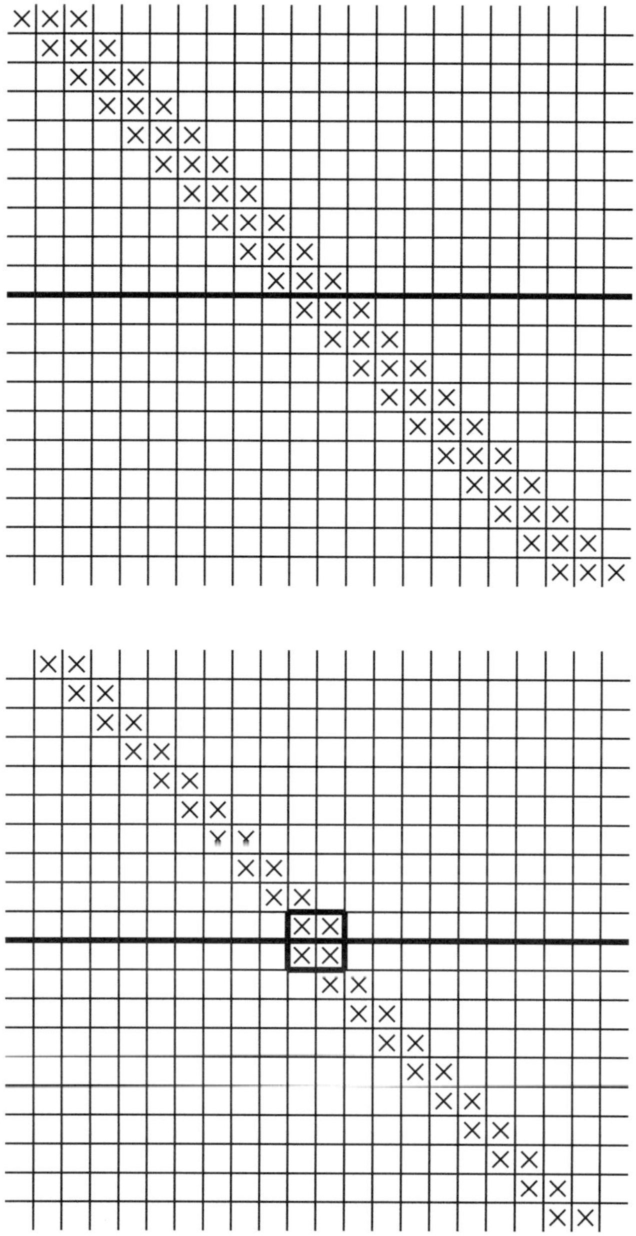

Fig. 2.7 An elimination algorithm that is more efficient than Gaussian elimination for some problems involving a sloping interface. The upper frame shows part of a tridiagonal matrix before elimination, with the dark horizontal line corresponding to the location of the interface. The lower frame shows the matrix after the elimination steps. Entries below the main diagonal are eliminated during a downward sweep to the interface. Entries above the main diagonal are eliminated during an upward sweep to the interface. After solving the 2×2 system in the dark box, the final steps are back substitution sweeping away from the interface in both directions

$$R_n(X) = \exp\left(ik_0\Delta x\right)\prod_{j=1}^{n}\frac{1+\gamma_{j,n}X}{1+\beta_{j,n}X} \cong \exp\left(ik_0\Delta x(1+X)^{1/2}\right). \tag{2.29}$$

The split-step Padé solution has the same form (but with different coefficients) as the split-step finite-difference solution in Eq. (2.20), which typically requires range steps that are a fraction of a wavelength. An advantage of approximations of the exponential function is that they provide higher-order accuracy in both the asymptotic parameter X and the numerical parameter $k_0\Delta x$, which often permits range steps of several wavelengths. A recursion formula in the Appendix is useful for computing the coefficients of $R_n(X)$. Since the operator on the right side of Eq. (2.29) is exponentially small for $X < -1$, an effective stability constraint is to require $R_n(X)$ to vanish at some point on this interval. With the sum form of $R_n(X)$, which is obtained with the method of partial fractions, efficiency can be maximized on a parallel-processing computer by assigning each term to a separate processor. The product form, which provides greater numerical stability, should be used on a single-processor computer. The split-step solutions are compared in Fig. 2.8 for a problem involving a 25 Hz source at $z = 285$ m in a 300 m deep ocean in which $c = 1500$ m/s that overlies a sediment in which $c = 1700$ m/s, $\rho = 1.5$ g/cm^3, and $\sigma = 0.5$. The split-step finite-difference solution was obtained using $n = 2$ and $\Delta r = 20$ m. The split-step Padé solution was obtained using $n = 6$ and $\Delta r = 500$ m. Accuracy starts to degrade for range steps larger than these values. For these inputs, the ratio of run times is about 8.3 on a single-processor computer and 25 on a parallel-processing computer. The reference solution was obtained using the split-step Padé solution with a small range step.

2.5 The Self-Starter

The self-starter is an approach for generating an initial condition (or starting field), which is required before marching the solution in range. The name of this approach reflects the fact that it is based on the parabolic equation itself. Placing a source term on the right side of Eq. (1.46), we obtain

$$\frac{\partial^2 p}{\partial x^2} + k_0^2(1+X)p = -2\tau\delta(x)\delta(z-z_0), \tag{2.30}$$

where $\tau = \sqrt{-2\pi i k(z_0)}$. Integrating Eq. (2.30) over an arbitrarily small interval about $x = 0$ and using the properties of the delta function and the symmetry of the solution, we obtain

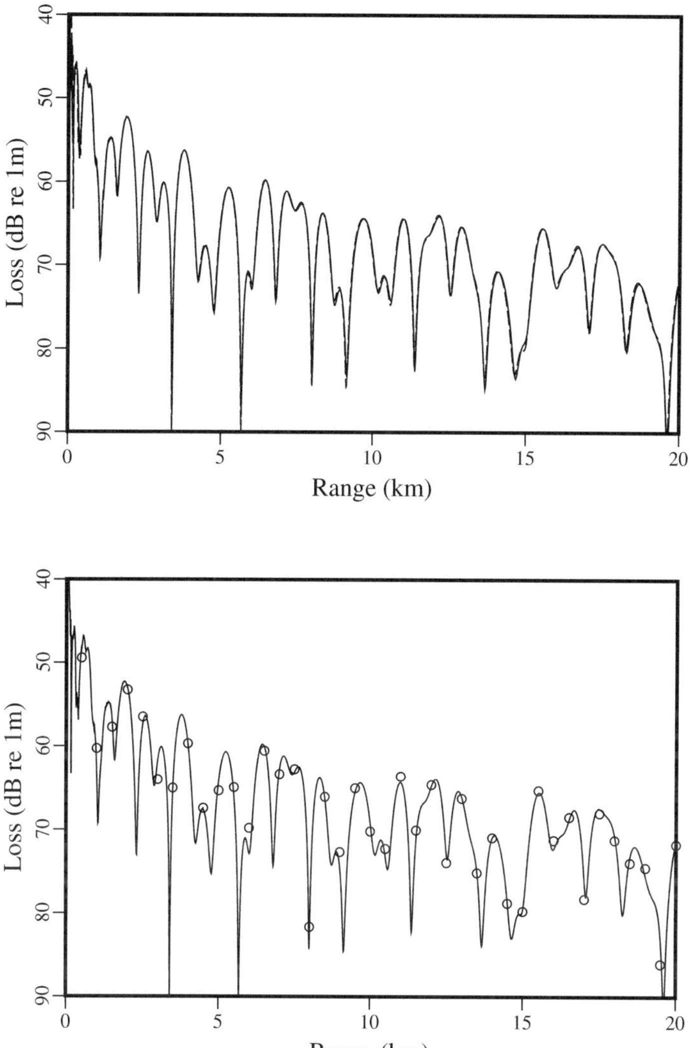

Fig. 2.8 Transmission loss at $z = 100$ m for a problem involving a 300 m deep water column overlying a sediment in which $c = 1700$ m/s, $\rho = 1.5$ g/cm^3, and $\sigma = 0.5$. In the top frame, the split-step finite-difference solution (solid curve) is in agreement with the reference solution (dashed curve). In the bottom frame, the split-step Padé solution (circles), which is sparsely sampled due to the large range step, is in agreement with the reference solution (solid curve)

$$\lim_{x \to 0^+} \frac{\partial p}{\partial x} = -\tau\delta(z - z_0). \tag{2.31}$$

Substituting Eq. (1.62) into Eq. (2.31) to eliminate the range derivative, we obtain

$$k_0(1+X)^{1/2}p(0,z) = i\tau\delta(z-z_0). \tag{2.32}$$

Since the solution of Eq. (2.32) is singular at $z = z_0$, we evaluate the initial condition away from the origin at $x = x_0$. From Eqs. (2.27) and (2.32), we obtain the self-starter [2],

$$p(x_0,z) = i\tau k_0^{-1}(1+X)^{-1/2}\exp\left(ik_0x_0(1+X)^{1/2}\right)\delta(z-z_0). \tag{2.33}$$

This equation can be derived directly using Eqs. (1.32), (1.35), and (1.58). For the point source case, we apply Eqs. (1.32), (1.39), and (1.58) to obtain

$$p(r_0,z) = \left(\frac{2\pi i}{k_0r_0}\right)^{1/2}(1+X)^{-1/4}\exp\left(ik_0r_0(1+X)^{1/2}\right)\delta(z-z_0). \tag{2.34}$$

Substituting the definition of τ into Eq. (2.33) for the case $k_0 = k(z_0)$, we obtain

$$p(x_0,z) = \left(\frac{2\pi i}{k_0}\right)^{1/2}(1+X)^{-1/2}\exp\left(ik_0x_0(1+X)^{1/2}\right)\delta(z-z_0), \tag{2.35}$$

and note the similarity with the point source solution in Eq. (2.34).

Since the application of the inverse of a differential operator increases smoothness, the stability of the self-starter can be improved by introducing the intermediate solution,

$$\xi(z) = (1-\varepsilon X)^{-2}\delta(z-z_0), \tag{2.36}$$

where the constant ε is chosen so that the inverse operator is well behaved. Applying a rational approximation, we obtain

$$p(x_0,z) = i\tau k_0^{-1}R_n(X)\xi(z), \tag{2.37}$$

$$\begin{aligned} R_n(X) &= \exp(ik_0x_0)\prod_{j=1}^{n}\frac{1+\gamma_{j,n}X}{1+\beta_{j,n}X} \\ &\cong (1-\varepsilon X)^2(1+X)^{-1/2}\exp\left(ik_0x_0(1+X)^{1/2}\right). \end{aligned} \tag{2.38}$$

For the point source case, we obtain

$$p(r_0,z) = \left(\frac{2\pi i}{k_0r_0}\right)^{1/2}R_n(X)\xi(z), \tag{2.39}$$

$$R_n(X) \cong (1 - \varepsilon X)^2 (1 + X)^{-1/4} \exp\left(ik_0 r_0 (1 + X)^{1/2}\right). \tag{2.40}$$

A recursion formula in the Appendix is useful for computing the coefficients of $R_n(X)$. It is natural to consider the choice $\varepsilon = -1$ since the operator on the right side of Eq. (2.33) is a function of $1 + X$. The original implementation of the self-starter [2] was based on this choice, which was later found to be problematic since the inverse operator is nearly singular for some problems [20]. When the attenuation vanishes, one of the eigenvalues vanishes at certain frequencies and the solution of Eq. (2.36) does not exist. In a homogeneous waveguide, for example, this occurs when H is a multiple of the wavelength. If the attenuation is weak, an eigenvalue may nearly vanish and cause the solution to be unstable when $\varepsilon = -1$. The choice $\varepsilon = i$ is effective for many cases.

The self-starter is identical to the spectral solution with the exception of the choices for the wave number samples and their weights in the sum. We compare these approaches by deriving the self-starter from the spectral solution for the plane geometry case. From Eqs. (1.42) and (1.43), we obtain

$$p(x_0, z) = \frac{\tau}{\pi} \int_{-\infty}^{\infty} \sum_m \frac{\zeta_m(z_0)\zeta_m(z)}{h^2 - k_m^2} \exp(ihx_0)dh. \tag{2.41}$$

From Eqs. (1.32) and (2.41), we obtain

$$p(x_0, z) = f(X)\delta(z - z_0), \tag{2.42}$$

$$f(X) = \frac{\tau}{\pi} \int_{-\infty}^{\infty} \frac{\exp(ihx_0)}{h^2 - k_0^2(1 + X)} dh. \tag{2.43}$$

Evaluating the integral in Eq. (2.43) in terms of the residue at $h = k_0(1 + X)^{1/2}$, we obtain the operator in Eq. (2.33) and conclude that the self-starter is equivalent to evaluating the spectral integral analytically and then fitting a rational function to the exact solution. The spectral solution is usually implemented by evaluating the spectral integral numerically. In some cases, the analytic implementation provides greater efficiency when the solution is desired at just one range [21].

2.6 Accurately Handling Range Dependence

Solving range-dependent problems is the most important application of the parabolic equation method. Although the factorization in Eq. (1.48) is not exact for range-dependent problems, Eq. (1.50) is a valid approximation in the limit of gradual range dependence. We approximate a range-dependent medium by a sequence of range-independent regions. With this approach, a sloping interface is approximated in terms of a series of stair steps. Since the factorization is exact in range-independent regions, the parabolic equation can be used to march the field through each region. To completely define the solution, it is necessary to specify a condition at the vertical interfaces between regions. Although approximate solutions may be obtained by conserving p, this approach often has significant errors [22]. Accurate solutions can be obtained for many problems by applying a single-scattering [23] or energy-conservation [24, 25] correction at the vertical interfaces. Sloping interfaces may be handled by working in a rotated coordinate system [26, 27]. Variable topography may be handled with a simple numerical approach [28], which may be used along with a mapping approach to handle sloping interfaces [29].

2.6.1 Single Scattering

The single-scattering correction is based on decomposing the field into incoming and outgoing fields and using the parabolic equation to simplify conditions on vertical interfaces. At a vertical interface between regions A and B, the incident p_i, transmitted p_t, and reflected p_r fields satisfy the following conditions for conservation of stress and normal displacement,

$$p_t = p_i + p_r, \tag{2.44}$$

$$\frac{1}{\rho_B}\frac{\partial p_t}{\partial x} = \frac{1}{\rho_A}\frac{\partial}{\partial x}(p_i + p_r). \tag{2.45}$$

If these conditions were identities, it would be possible to substitute Eq. (2.44) into Eq. (2.45) and eliminate p_r. Since the conditions hold only on the vertical interface, we apply Eqs. (1.50) and (1.51) to eliminate the range derivatives in Eq. (2.45) and obtain

$$\rho_B^{-1}(1+X_B)^{1/2}p_t = \rho_A^{-1}(1+X_A)^{1/2}(p_i - p_r). \tag{2.46}$$

A negative sign appears in front of p_r, which propagates in the negative x direction. Inverting the operator on the left side of Eq. (2.46), we obtain

$$p_t = (1+X_B)^{-1/2}\rho_B\rho_A^{-1}(1+X_A)^{1/2}(p_i - p_r). \tag{2.47}$$

Taking the sum of Eqs. (2.44) and (2.47), we obtain

$$\begin{aligned} p_t = {} & \frac{1}{2}p_i + \frac{1}{2}(1+X_B)^{-1/2}\rho_B\rho_A^{-1}(1+X_A)^{1/2}p_i + \\ & \frac{1}{2}\left[1 - (1+X_B)^{-1/2}\rho_B\rho_A^{-1}(1+X_A)^{1/2}\right]p_r. \end{aligned} \tag{2.48}$$

Based on the assumption of gradual range dependence, the reflected field should be dominated by the incident field and the operator in the brackets should be small. Neglecting the term in Eq. (2.48) that is small to second order, we obtain the single-scattering correction,

$$p_t = \frac{1}{2}p_i + \frac{1}{2}(1+X_B)^{-1/2}\rho_B\rho_A^{-1}(1+X_A)^{1/2}p_i. \tag{2.49}$$

The exact single-scattering solution conserves the stress and the normal displacement across the vertical interface. Since the first term in Eq. (2.49) corresponds to conservation of stress and the second term corresponds to conservation of normal displacement, the approximate single-scattering solution conserves these quantities in an average sense.

2.6.2 *Energy Conservation*

For many range-dependent problems, accurate solutions may be obtained by applying a condition for conserving energy flux across the vertical interfaces between the range-independent regions that approximate a range-dependent environment. We derive this condition for the case in which the attenuation vanishes. The energy flux is proportional to

$$E = \mathrm{Im}\int_0^H \rho^{-1}p^*\frac{\partial p}{\partial x}dz. \tag{2.50}$$

A direct implementation of Eq. (2.50) would be a nonlinear interface condition. A linear condition can be obtained using the modal representation. Substituting Eq. (1.24) into Eq. (2.50), we obtain

$$E = \mathrm{Im}\int_0^H \rho^{-1}\left(\sum_m p_m^*(x)\zeta_m(z)\right)\left(\sum_m ik_m p_m(x)\zeta_m(z)\right)dz. \tag{2.51}$$

From Eq. (1.26), we obtain

$$E = \mathrm{Im}\int_0^H \left(\rho^{-1/2}\sum_m k_m^{1/2}p_m^*(x)\zeta_m(z)\right)\left(\rho^{-1/2}\sum_m k_m^{1/2}p_m(x)\zeta_m(z)\right)dz. \tag{2.52}$$

Applying Eq. (1.58), we obtain

$$E = k_0 \int_0^H \left(\rho^{-1/2} \sum_m p_m^*(x)(1+X)^{1/4} \zeta_m(z) \right) \times \left(\rho^{-1/2} \sum_m p_m(x)(1+X)^{1/4} \zeta_m(z) \right) dz. \tag{2.53}$$

Moving the operator outside the sums, we obtain

$$E = k_0 \int_0^H \left| \rho^{-1/2}(1+X)^{1/4} p \right|^2 dz. \tag{2.54}$$

At a vertical interface between regions A and B, we obtain the energy-conservation condition,

$$\rho_B^{-1/2}(1+X_B)^{1/4} p_t = \rho_A^{-1/2}(1+X_A)^{1/4} p_i. \tag{2.55}$$

This condition is related to the self-starter for a point source, which also involves the fourth root of the operator. The incident and transmitted fields are identified with arrays of point sources with appropriate amplitudes. Since p is continuous across horizontal interfaces but not necessarily across vertical interfaces in this solution, the energy-conservation condition can give rise to Gibbs oscillations, which may be annihilated by using a stabilized rational approximation to propagate the solution. To leading order in the limit of small propagation angles, Eq. (2.55) reduces to

$$k_B^{1/2} \rho_B^{-1/2} p_t = k_A^{1/2} \rho_A^{-1/2} p_i. \tag{2.56}$$

For a problem with a sloping interface, this condition is applied at only one grid point on each vertical interface (the grid points that correspond to the rises of the stair steps), and it often provides accurate solutions [25].

The examples appearing in Fig. 2.9 are useful for testing the single-scattering and energy-conserving solutions. Both problems involve a 25 Hz source in a water column in which $c = 1500$ m/s and an upward sloping ocean bottom. For the case in which there is a relatively low contrast across the ocean bottom interface, the sediment parameters are $c = 1700$ m/s, $\rho = 1.5$ g/cm^3, and $\sigma = 0.5$, and the bathymetry decreases linearly from 200 m at the source to zero at $r = 4$ km. For the high contrast case, the sediment parameters are $c = 3400$ m/s, $\rho = 2.5$ g/cm^3, and $\sigma = 0.5$, and the bathymetry decreases linearly from 400 m at $r = 3$ km to 200 m at $r = 7$ km. A reference solution is obtained for the low contrast case with the energy-conserving solution, which is known to be accurate for this problem [25], and a rational approximation generated with one stability constraint. A finite-element model [30] is used to obtain a reference solution for the high contrast case. The solutions appearing in Fig. 2.10 for the low contrast case illustrate an amplitude error in the solution that conserves p across vertical interfaces and Gibbs oscillations in an

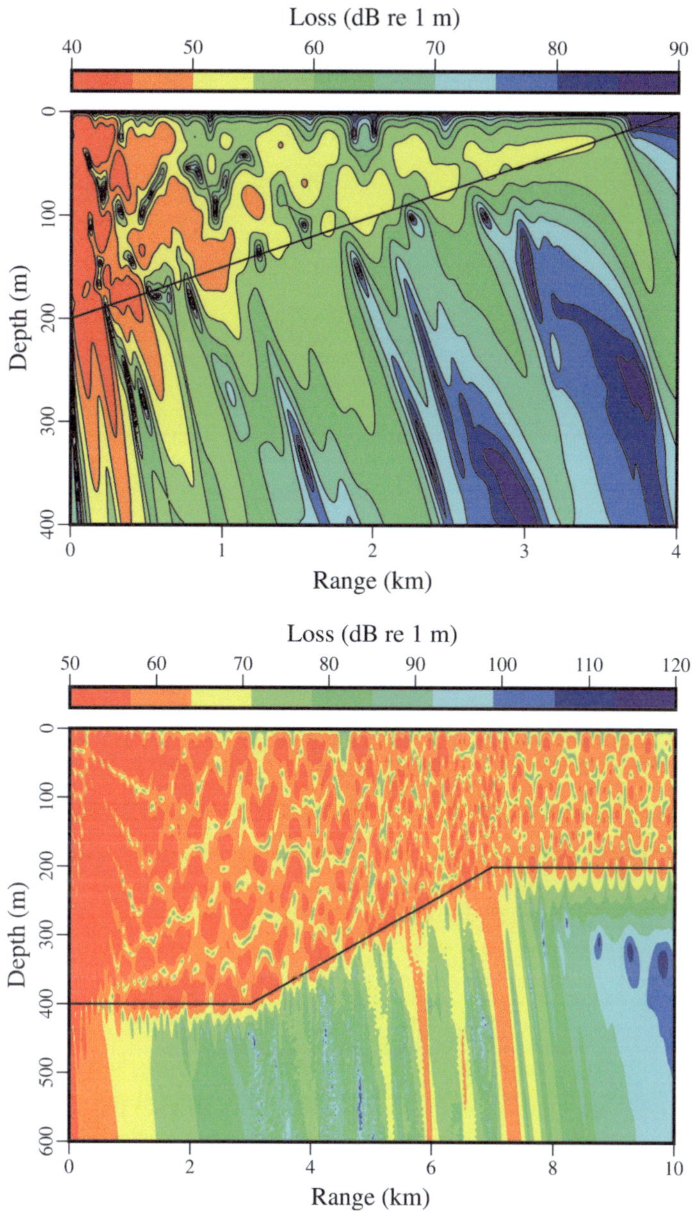

Fig. 2.9 Range-dependent test cases involving upward sloping ocean bottoms. For the low contrast case (top), the sediment parameters are $c = 1700$ m/s, $\rho = 1.5$ g/cm^3, and $\sigma = 0.5$. For the high contrast case (bottom), the sediment parameters are $c = 3400$ m/s, $\rho = 2.5$ g/cm^3, and $\sigma = 0.5$

energy-conserving solution that was generated using Padé coefficients (no stability constraints). The solutions appearing in Fig. 2.11 for the high contrast case illustrate

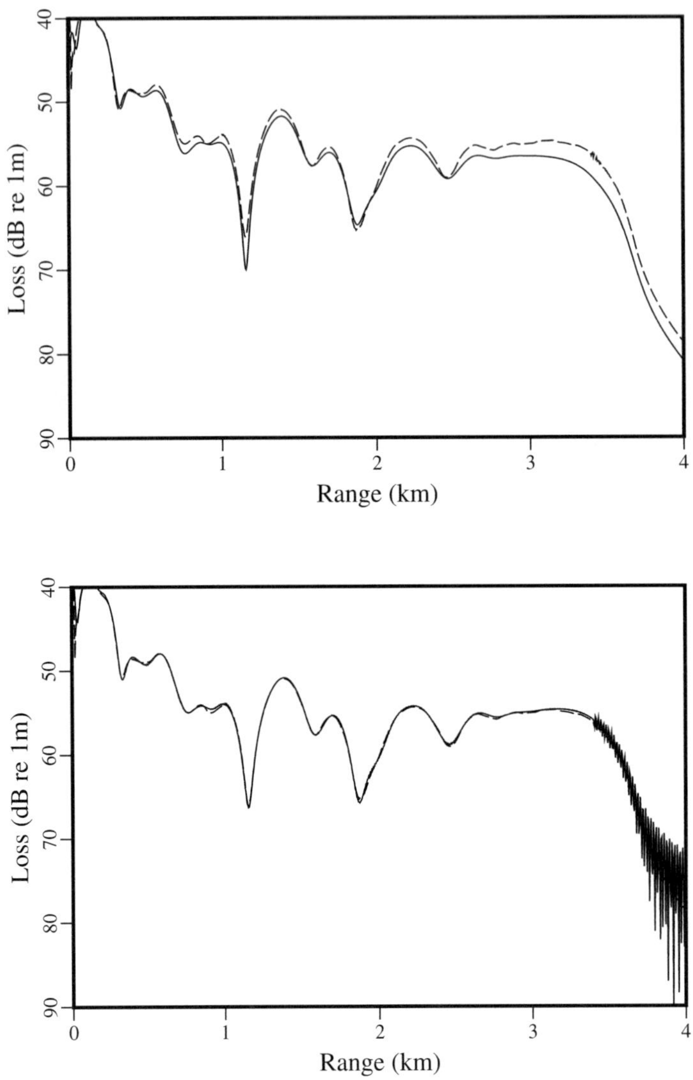

Fig. 2.10 Transmission loss at $z = 30$ m for the low contrast test case. The dashed curves were obtained using the energy-conserving solution and rational approximations that were generated using one stability constraint. The solid curves illustrate an amplitude error in the solution that conserves p across vertical interfaces (top) and Gibbs oscillations in the energy-conserving solution that was generated using Padé coefficients (bottom)

an amplitude error in the solution that conserves p across vertical interfaces and the accuracy of the energy-conserving solution. The solutions appearing in Fig. 2.12 illustrate that the single-scattering solution is accurate for the low contrast case but has an amplitude error for the high contrast case.

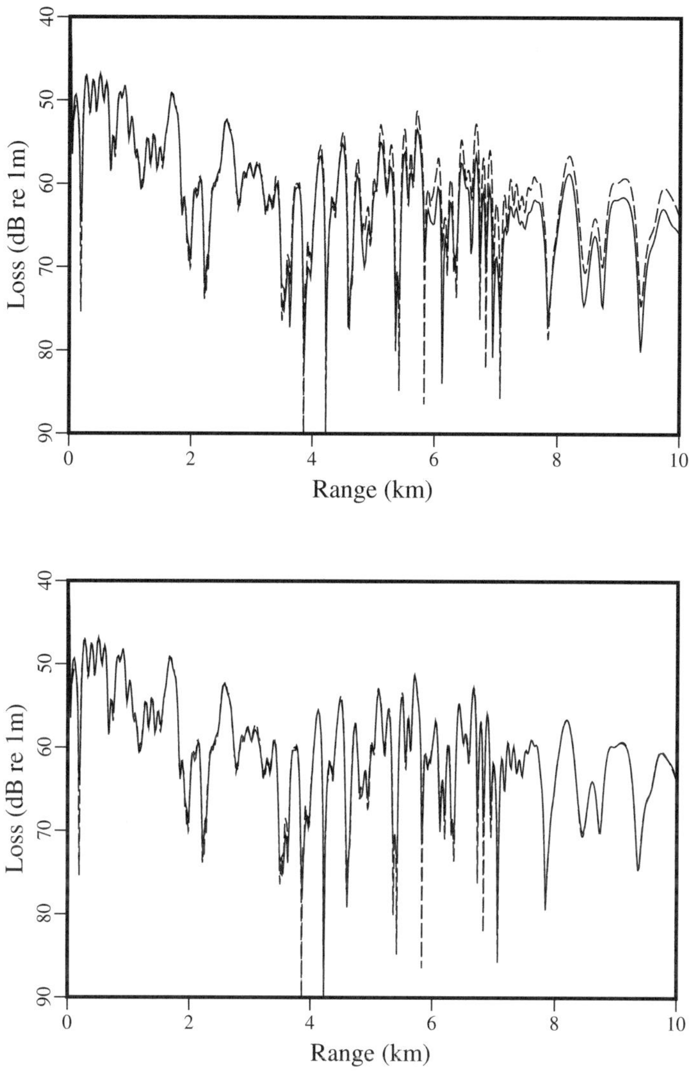

Fig. 2.11 Transmission loss at $z = 180$ m for the high contrast test case. The dashed curves are reference solutions. The solid curves illustrate an amplitude error in the solution that conserves p across vertical interfaces (top) and the accuracy of the energy-conserving solution (bottom)

2.6.3 *Rotated Coordinates, Variable Topography, and Mapping*

For many years, range-dependent problems were solved with the parabolic equation method by simply updating the entries of the propagation matrices in order to account for the range dependence of the sound speed and density. With this

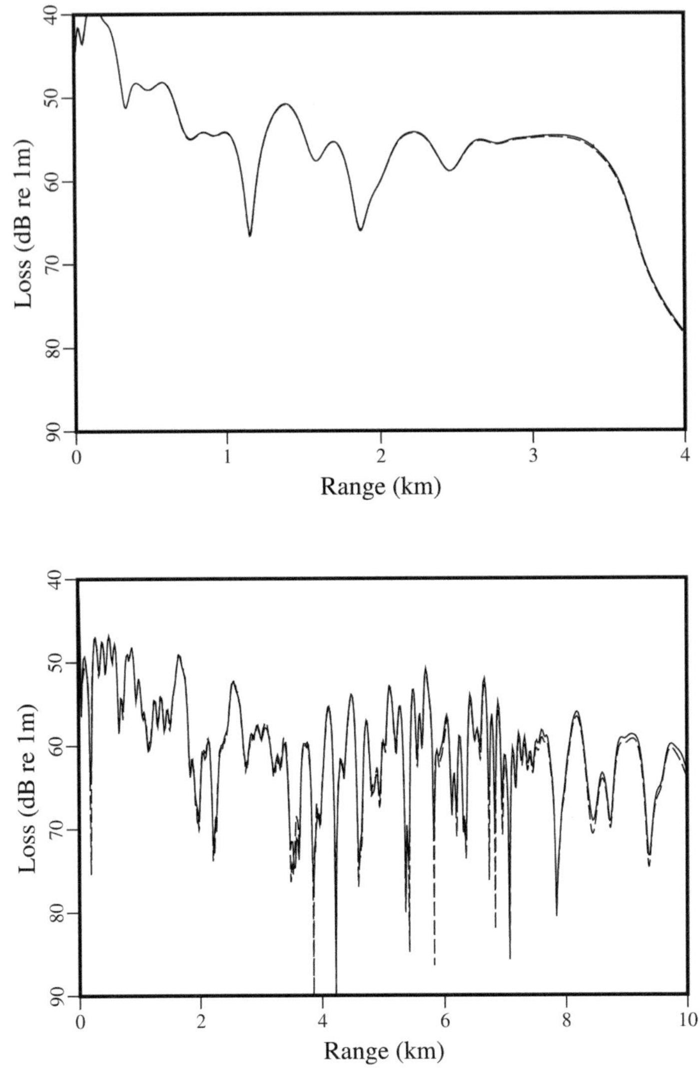

Fig. 2.12 Transmission loss at $z = 30$ m for the low contrast test case (top) and at $z = 180$ m for the high contrast test case (bottom). The dashed curves are reference solutions. The single-scattering solution (solid curves) is accurate for the low contrast case but has an amplitude error for the high contrast case

approach, the dependent variable is conserved across the vertical interfaces between range-independent regions. Since p is conserved across an interface in the exact two-way solution, it seemed reasonable to conserve this quantity in an approximate one-way solution, but accuracy issues came to light after a series of range-dependent benchmark problems were proposed [22]. It was initially believed that the errors were due to the neglect of back-scattered energy, but this misconception was

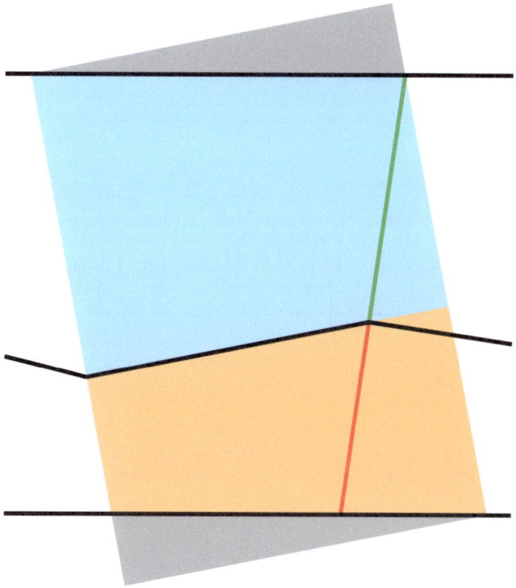

Fig. 2.13 Geometry of the variable rotated parabolic equation solution. The horizontal black lines correspond to the ocean surface and the bottom of the computational grid. The sloping black line segments correspond to an ocean-sediment interface that is approximated in terms of a series of constant slopes. On the colored line that is orthogonal to the interface, an initial condition is obtained by interpolation (red) and extrapolation (green). The field is extrapolated beyond the change in slope by overshooting a short distance while keeping the slope constant, as indicated by the blue and tan regions

resolved when accurate solutions were constructed from outgoing rays [31]. These findings motivated the development of improvements to the accuracy of parabolic equation solutions for range-dependent problems. The first progress in that direction was the rotated parabolic equation [26], which handles a sloping interface by working in a rotated coordinate system in which one of the axes (the preferred direction) is parallel to the interface. A horizontal boundary in the original coordinates becomes a sloping boundary in rotated coordinates, but the condition $p = 0$ may be imposed at a sloping boundary simply by adding and subtracting grid points as the location of the boundary changes. As illustrated in Fig. 2.13, the rotated parabolic equation may be applied to problems involving variable slope by propagating the solution slightly beyond the range at which a change in slope occurs. An initial condition is then obtained at the beginning of the next region by interpolation. Variable rotated parabolic equation solutions appear in Fig. 2.14 for the high contrast problem in Fig. 2.9 and for a modified version of that problem in which $c = 1700$ m/s and $\rho = 1.5$ g/cm^3 in the sediment. The variable rotated parabolic equation solutions are accurate for both cases.

Sloping interfaces may also be handled with a mapping approach [29] that is based on the transformation,

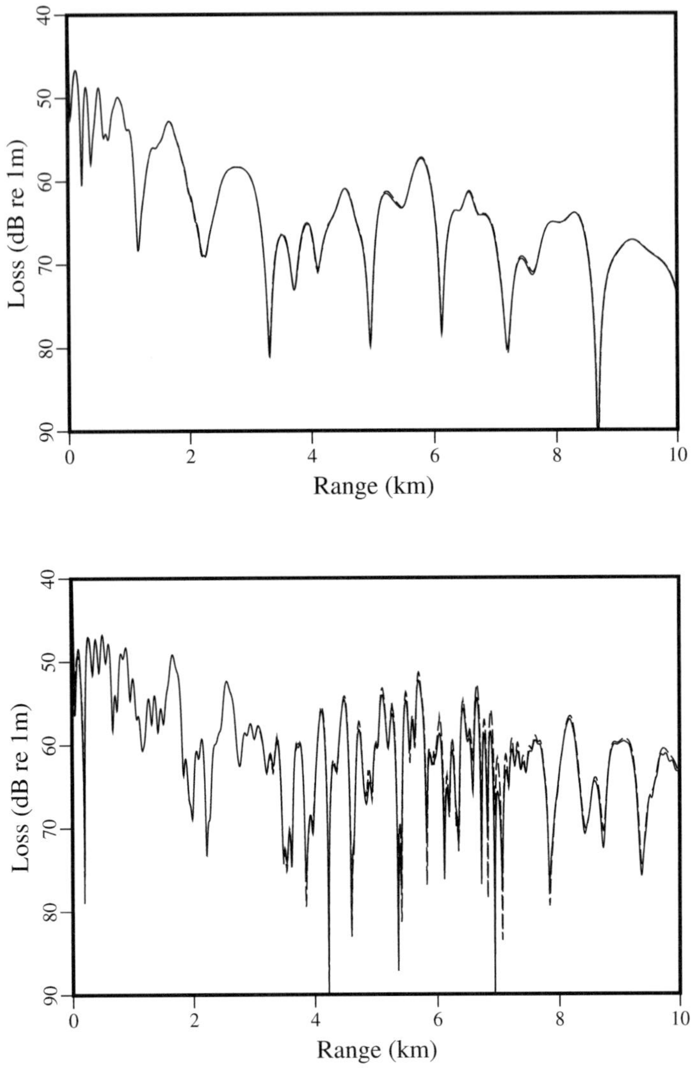

Fig. 2.14 Transmission loss at $z = 180$ m for modified (top) and unmodified (bottom) versions of the high contrast test case in Fig. 2.9. The dashed curves are reference solutions. The variable rotated parabolic equation solutions (solid curves) are accurate for both cases

$$\begin{pmatrix} \tilde{x} \\ \tilde{z} \end{pmatrix} = \begin{pmatrix} x \\ z - d(x) \end{pmatrix}, \tag{2.57}$$

where $z = d(x)$ is the depth of the interface. With this change of variables, a sloping interface becomes a flat interface at the expense of replacing a flat surface with a sloping surface and introducing new terms in the wave equation. The sloping surface can be handled using the same approach that is used in rotated coordinates, and the new terms can be neglected if the slope of the interface is sufficiently small. Applying the transformation to Eq. (1.14) and neglecting terms involving the slope, we obtain

$$\frac{\partial^2 p}{\partial \tilde{x}^2} + \rho \frac{\partial}{\partial \tilde{z}} \left(\frac{1}{\rho} \frac{\partial p}{\partial \tilde{z}} \right) + k^2 p = 0. \tag{2.58}$$

Appearing in Fig. 2.15 are mapped solutions for a problem involving a 10 Hz source at $z = 390$ m in a water column in which $c = 1500$ m/s [32]. In the sediment $c = 1700$ m/s, $\rho = 1.5$ g/cm^3, and $\sigma = 0.1$. The bathymetry decreases from 400 m at $r = 4$ km to 100 m at $r = r_{\text{max}}$. The mapped solution has small errors for the case $r_{\text{max}} = 20$ km but is nearly identical to the reference solution for the case $r_{\text{max}} = 80$ km. Appearing in Fig. 2.16 are solutions of the unmapped problem that conserve p across vertical interfaces. The lengths of the range-independent regions are the only differences between these cases, which involve the same sequences of vertical interfaces, and the error in the p conserving solution is about the same for both cases. When a range-dependent region is approximated in terms of a series of range-independent regions, the mapping vertically translates each region so that the runs are all at the same depth. The adiabatic mode solution is invariant under such translations, but the mapping solution is superior to the adiabatic mode solution for the mode cutoff problem that appears in Fig. 2.17. The mapping solution properly accounts for energy coupling into the beam that radiates into the sediment. The adiabatic mode solution does not account for such coupling effects.

The neglected terms can give rise to errors, which we illustrate for an example involving a 25 Hz source at $z = 400$ m in an ocean with the canonical sound speed profile [33],

$$c(z) = c_* \left(1 + \varepsilon \left(\frac{z - z_*}{L} + \exp \left(\frac{z - z_*}{L} \right) - 1 \right) \right), \tag{2.59}$$

where $c_* = 1500$ m/s, $L = 600$ m, $z_* = 1000$ m, and $\varepsilon = 0.007$. The bathymetry is 5 km for $r < 25$ km, 1 km for $r > 50$ km, and linearly sloping for 25 km $< r <$ 50 km. In the sediment, $c = 1550$ m/s, $\rho = 1.15$ g/cm^3, and $\sigma = 0.5$. The mapping solution appearing in Fig. 2.18 contains bends in the Lloyd's mirror beams at $r = 25$ km, which are artifacts of neglecting the slope terms. The beams are smooth in the plot of the solution in mapped coordinates appearing in Fig. 2.18. The bends are introduced when the solution is mapped back to the original coordinates. The error associated with the bends can be reduced by approximately introducing equal but opposite bends with the correction,

$$p(\tilde{x}, \tilde{z}) \rightarrow \exp\left(-ik_0 \tilde{z} \sin \delta\right) p(\tilde{x}, \tilde{z}), \tag{2.60}$$

each time the slope changes, where δ is the change in slope. Since the sloping interface is handled properly in mapped coordinates, energy is conserved and the

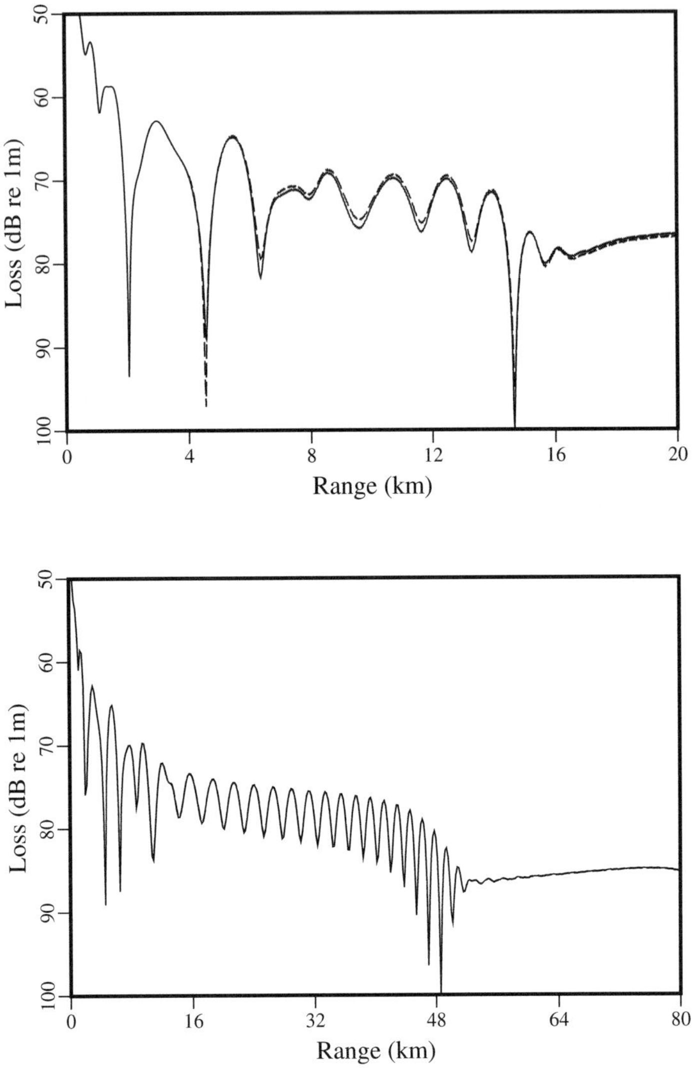

Fig. 2.15 Transmission loss at $z = 30$ m for problems in which the bathymetry decreases linearly from 400 m at $r = 4$ km to 100 m at $r = r_{\text{max}}$. The solid curves are mapping solutions. The dashed curves were obtained with the energy-conserving parabolic equation

correction in Eq. (2.60) affects the phase. In contrast, the correction in Eq. (2.56) affects the amplitude since the parabolic equation properly handles the phase. Appearing in Fig. 2.19 is the corrected mapping solution, which is in agreement with the energy-conserving solution. A quantitative test of the corrected mapping solution appears in Fig. 2.20. This problem involves a 25 Hz source at $z = 100$ m in the water column. In the sediment, $c = 1700$ m/s, $\rho = 1.5$ g/cm^3, and $\sigma = 0.5$. The bathymetry is 200 m at $r = 0$ and $r = 20$ km, 40 m at $r = 10$ km and $r = 30$ km, and

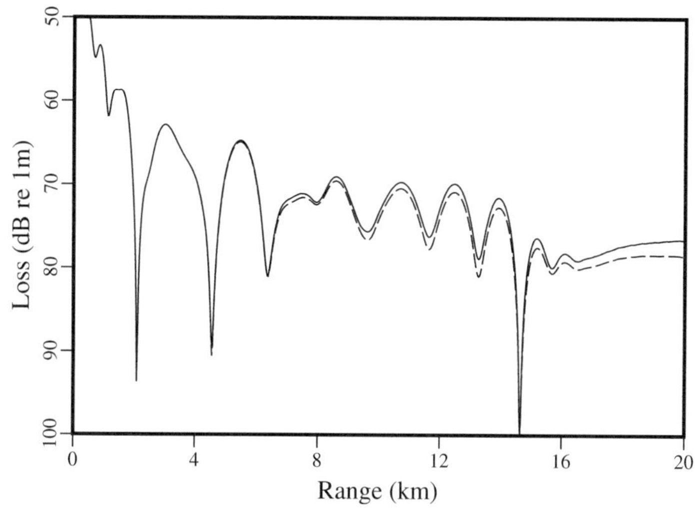

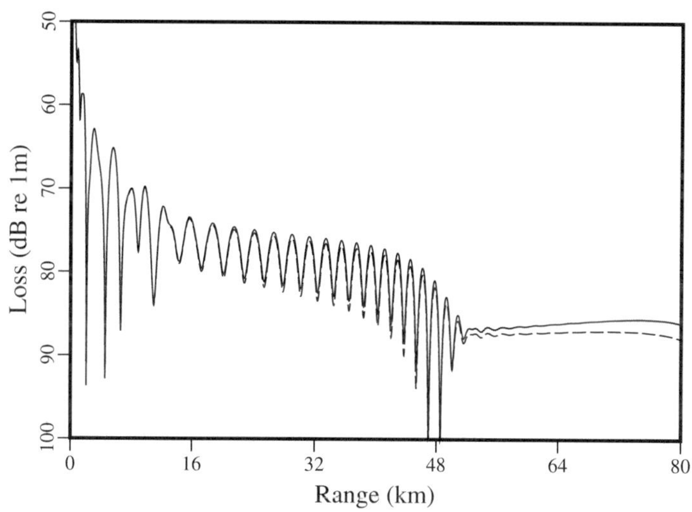

Fig. 2.16 Transmission loss at $z = 30$ m for problems in which the bathymetry decreases linearly from 400 m at $r = 4$ km to 100 m at $r = r_{\max}$. The solid curves conserve p across the vertical interfaces between range-independent regions. The dashed curves were obtained with the energy-conserving parabolic equation

linearly sloping between these points. The uncorrected mapping solution breaks down for $r > 10$ km. The corrected mapping solution is in agreement with the reference solution.

The approaches introduced in this section are applicable to problems involving variable topography [28, 29]. We illustrate this capability for a modified version of the mode cutoff problem. A 25 Hz source is located at $z = 112$ m in an ocean that is

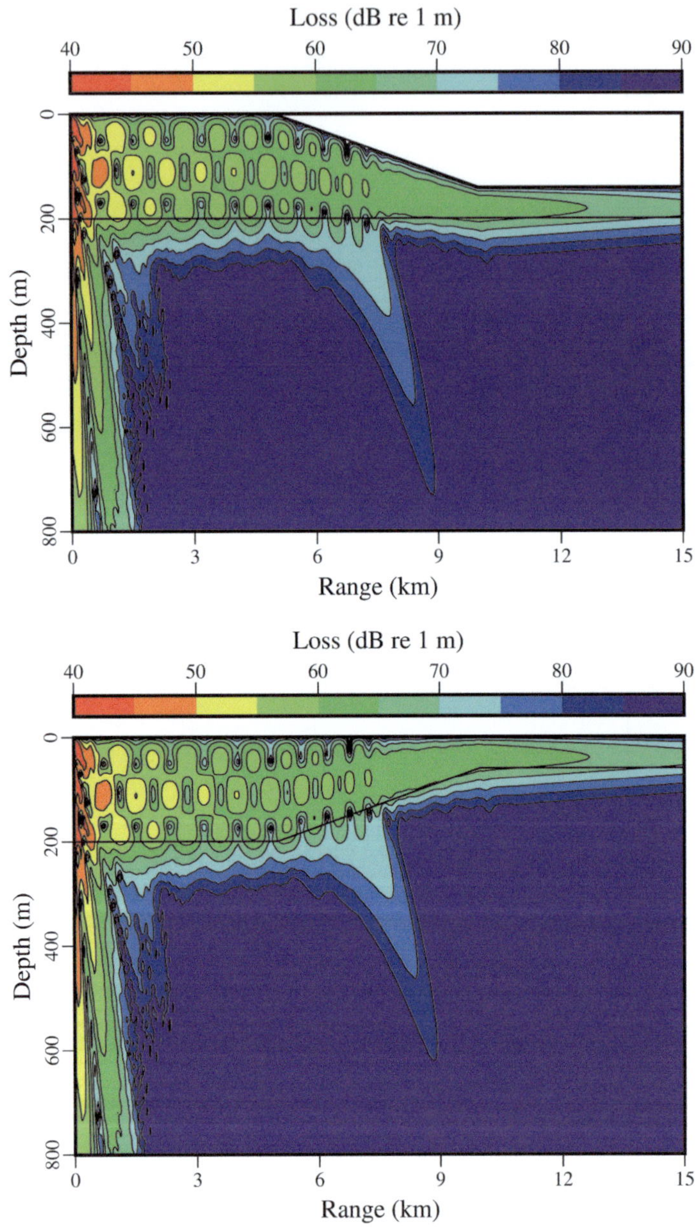

Fig. 2.17 Mapping solution of the mode cutoff problem appearing in Fig. 1.1. The problem that is solved in mapped coordinates (top) has a horizontal ocean-sediment interface at $z = 200$ m but variable topography. A solution of the original problem (bottom) is obtained by vertically translating the range-independent regions back into their original positions

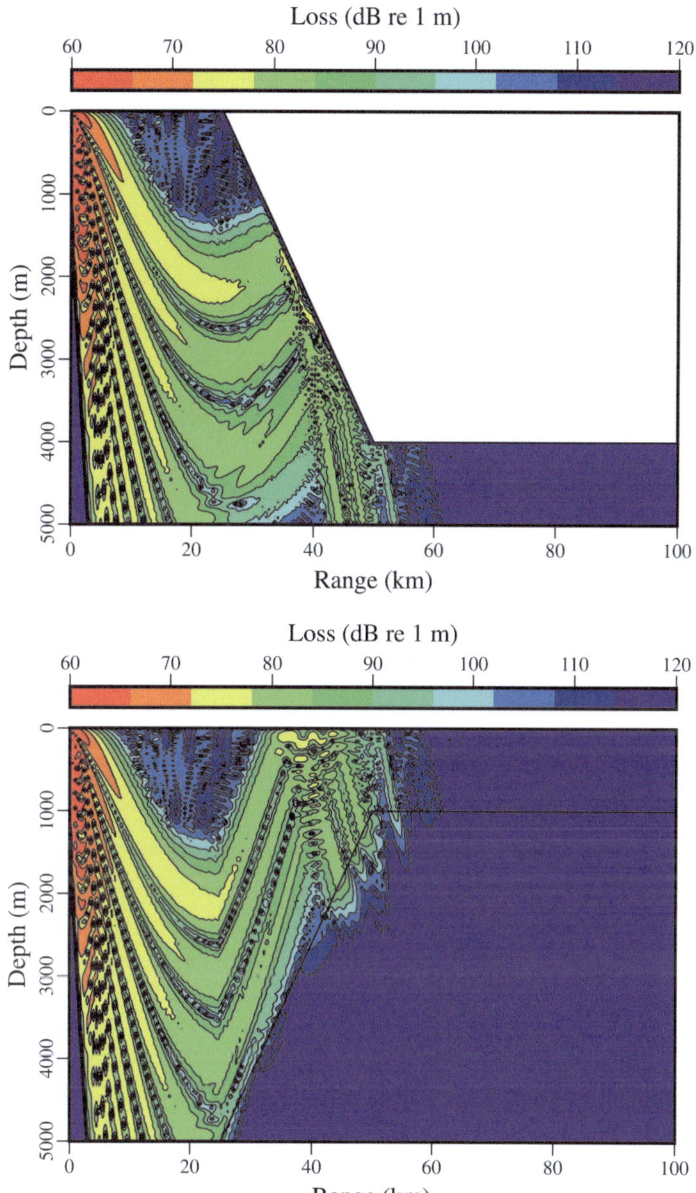

Fig. 2.18 Uncorrected mapping solution of a deep water problem. The Lloyd's mirror beams appear in the solution of the mapped problem (top), which has variable topography. After mapping back to the original coordinates (bottom), bends are introduced into the beams

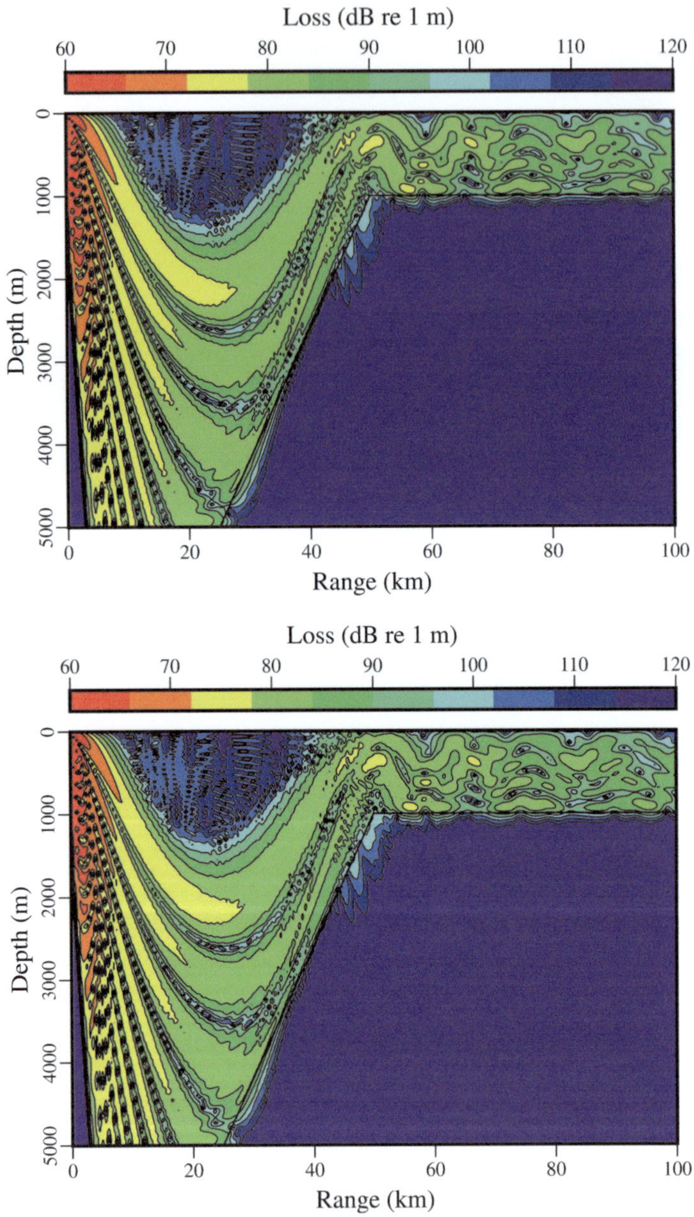

Fig. 2.19 The mapping solution (top) obtained using the correction in Eq. (2.60) is in agreement with the energy-conserving solution (bottom)

200 m deep for $r < 5$ km and linearly sloping to a beach at $r = 12.5$ km. Beyond the beach, the topography has the same slope and increases to 200 m above sea level at $r = 20$ km. In a 100 m thick sediment layer, $c = 1704.5$ m/s, $\rho = 1.15$ g/cm^3, and

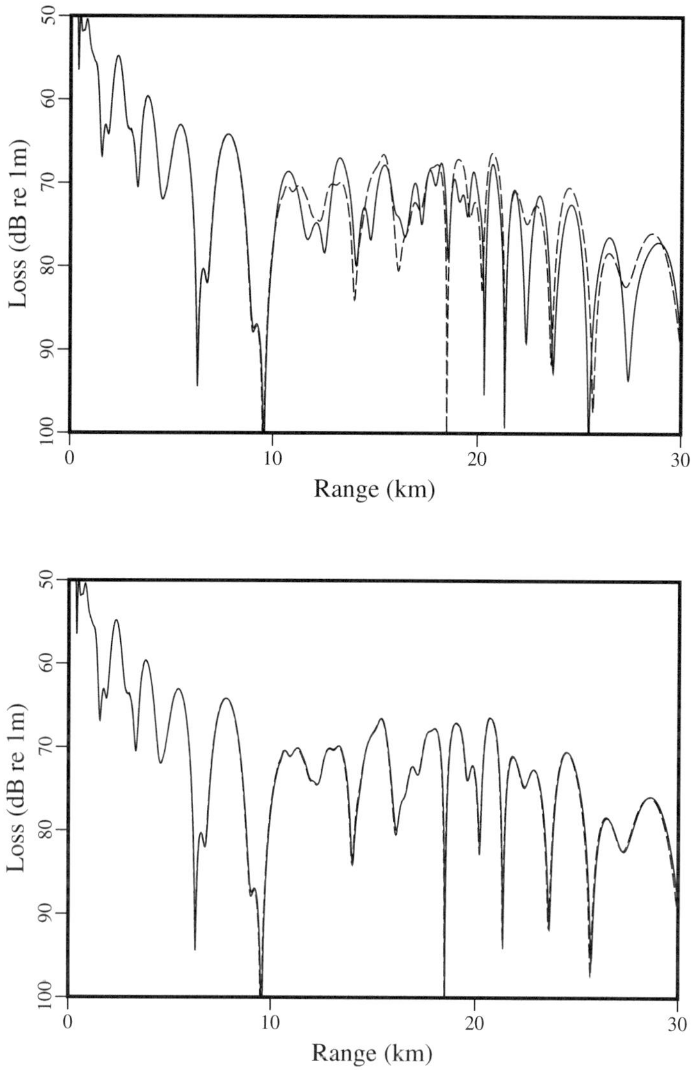

Fig. 2.20 Transmission loss at $z = 30$ m for a problem involving changes in the slope of the ocean-sediment interface at $r = 10$ km and $r = 20$ km. The dashed curves were obtained with the energy-conserving parabolic equation. The solid curves are the uncorrected (top) and corrected (bottom) mapping solutions

$\sigma = 0.1$. In the basement, $c = 1850$ m/s, $\rho = 1.5$ g/cm^3, and $\sigma = 0.25$. The solutions appearing in Fig. 2.21 were obtained using the mapping approach and the energy-conserving correction in the original coordinates.

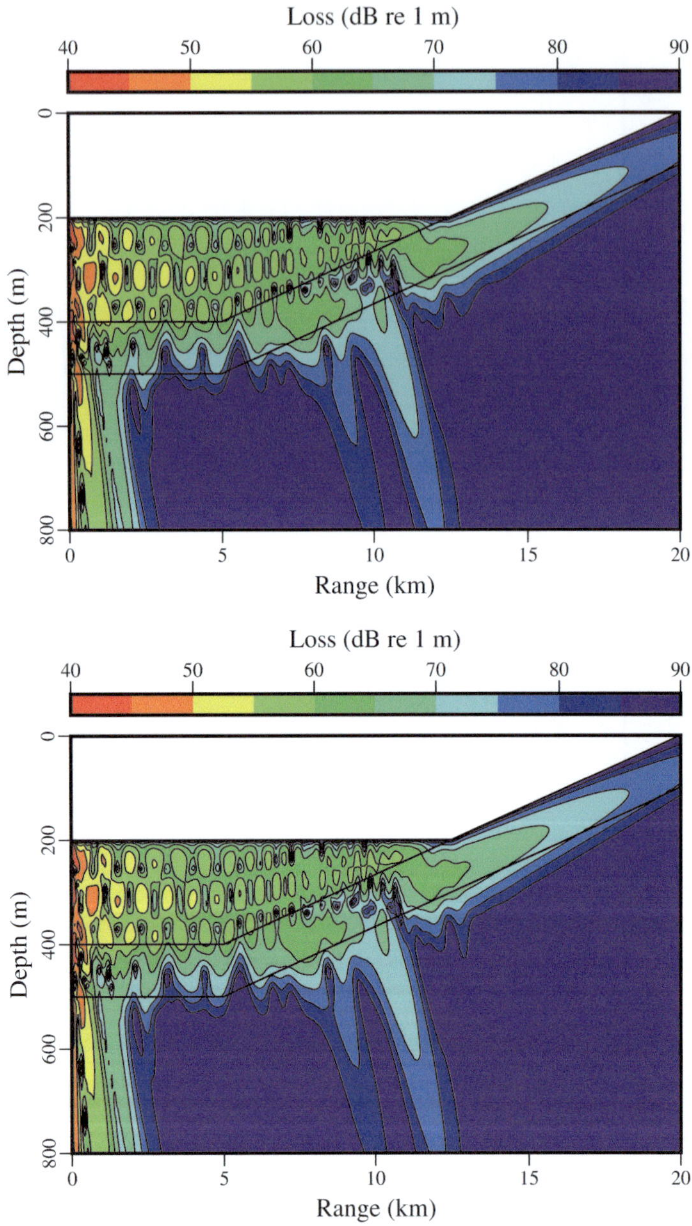

Fig. 2.21 Solutions of a problem involving propagation across a beach and variable topography beyond the beach. The mapping solution (top) was obtained by flattening the top boundary. The energy-conserving solution (bottom) was obtained by working in the original coordinates and introducing new grid points to account for varying topography

2.7 Three-Dimensional Problems

Azimuthal coupling can often be neglected but is significant for some problems. In ocean acoustics, this occurs when there are sufficiently large variations in bathymetry. Several parabolic equation techniques have been developed for three-dimensional problems. The three-dimensional parabolic equation is a direct generalization of the two-dimensional parabolic equation [4–6]. The adiabatic mode parabolic equation can be used to solve the horizontal wave equations for the mode coefficients of the adiabatic mode solution [7, 8]. This significantly improves the efficiency of three-dimensional calculations, which are difficult even with the parabolic equation method.

2.7.1 *Three-Dimensional Parabolic Equations*

The parabolic equation method is applicable to three-dimensional problems involving gradual horizontal variations. Including the azimuthal coupling term in Eq. (1.13), we obtain

$$\left(\frac{\partial^2}{\partial r^2} + k_0^2(1 + X + Y)\right)p = 0, \tag{2.61}$$

$$X = k_0^{-2}\left(\rho\frac{\partial}{\partial z}\frac{1}{\rho}\frac{\partial}{\partial z} + k^2 - k_0^2\right), \tag{2.62}$$

$$Y = \frac{1}{k_0^2 r^2}\frac{\partial^2}{\partial \theta^2}. \tag{2.63}$$

Factoring the operator in Eq. (2.61) and assuming that outgoing energy dominates back-scattered energy, we obtain

$$\frac{\partial p}{\partial r} = ik_0(1 + X + Y)^{1/2}p. \tag{2.64}$$

Substituting a linear Taylor approximation into Eq. (2.64), we obtain the narrow-angle parabolic equation,

$$\frac{\partial p}{\partial r} = ik_0\left(1 + \frac{1}{2}(X + Y)\right)p, \tag{2.65}$$

which can be solved with the splitting method. Substituting a rational-linear approximation into Eq. (2.64), we obtain the wide-angle parabolic equation,

$$\frac{\partial p}{\partial r} = ik_0 \left(1 + \frac{\frac{1}{2}(X+Y)}{1+\frac{1}{4}(X+Y)} \right) p, \tag{2.66}$$

which cannot be solved with the standard splitting method.

Applying a rational approximation that is correct to wide angles in depth and narrow angles in azimuth, we obtain

$$\frac{\partial p}{\partial r} = ik_0 \left(1 + \frac{\frac{1}{2}X}{1+\frac{1}{4}X} + \frac{1}{2}Y \right) p. \tag{2.67}$$

This equation contains a leading-order correction to Eq. (1.53) when azimuthal coupling is treated as a perturbation, and it may be solved with the splitting method, which requires solutions of the equations,

$$\frac{\partial p}{\partial r} = ik_0 \left(1 + \frac{\frac{1}{2}X}{1+\frac{1}{4}X} \right) p, \tag{2.68}$$

$$\frac{\partial p}{\partial r} = \frac{1}{2} ik_0 Y p. \tag{2.69}$$

It may be possible to improve stability by replacing the one-term Padé approximation in Eq. (2.68) with a multi-term rational approximation that annihilates the evanescent modes. With this approach, accuracy may also be improved if higher-order terms in X are more important than the leading-order cross term XY. The matrices involved in the solution of Eq. (2.69) are tridiagonal but have additional entries in the upper right and lower left corners to account for the periodic condition $p(\theta + 2\pi) = p(\theta)$. To illustrate azimuthal coupling, we consider a problem involving a 25 Hz source at $z = 25$ m in the water column, a sediment in which $c = 1700$ m/s, $\rho = 1.5$ g/cm^3, and $\sigma = 0.5$, and the bathymetry [6],

$$d(x,y) = \left[3 - \sin\left(\frac{2\pi x}{6\ \text{km}} \right) \right] 50\ \text{m}. \tag{2.70}$$

As shown in Fig. 2.22, rays get trapped in one of the valleys that parallel the y axis. In Fig. 2.23, this horizontal ducting effect is evident in a comparison of solutions that were obtained with and without the azimuthal coupling term. These solutions are compared quantitatively in Fig. 2.24.

2.7.2 *Adiabatic Mode Solution*

Computation times may be reduced by several orders of magnitude in going from Eq. (1.10) to Eq. (2.67). An additional large gain in efficiency can be achieved by

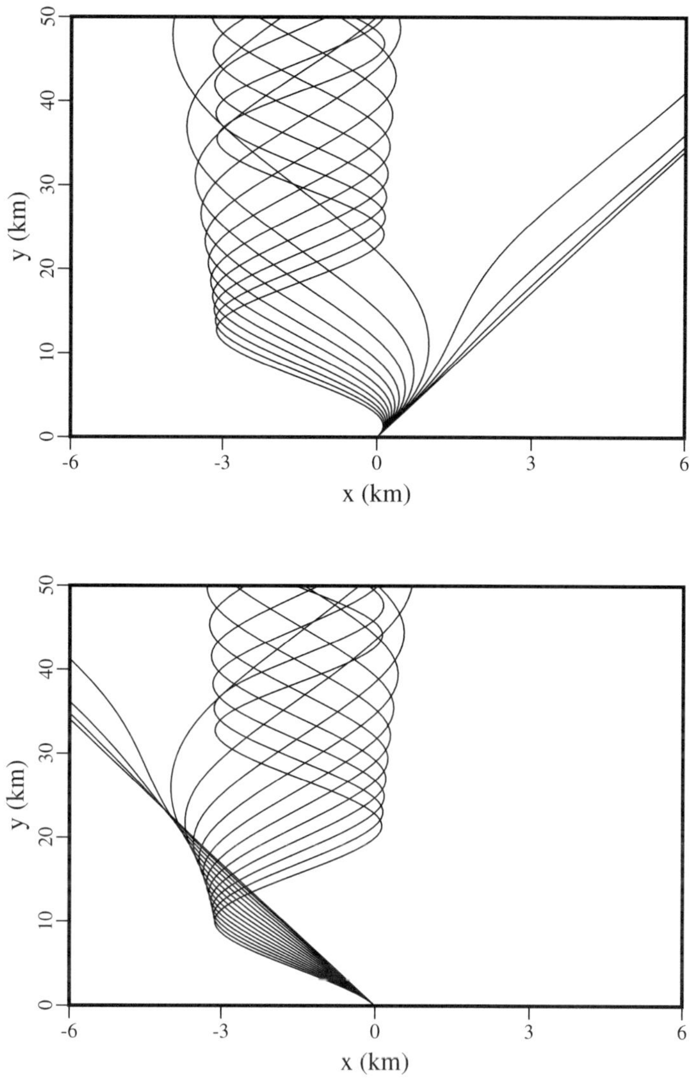

Fig. 2.22 Paths of rays launched from the origin at 15 equally spaced vertical angles up to 30° and horizontal angles of 80° (top) and 100° (bottom) from the x axis in an environment in which the bathymetry is a sinusoidal function of x. Many of the rays are trapped within one of the valleys as they repeatedly reflect from the ocean surface and ocean-sediment interface

using a local normal mode expansion and solving horizontal wave equations for the mode coefficients with the parabolic equation method. Including the azimuthal coupling term in Eq. (1.13), we obtain

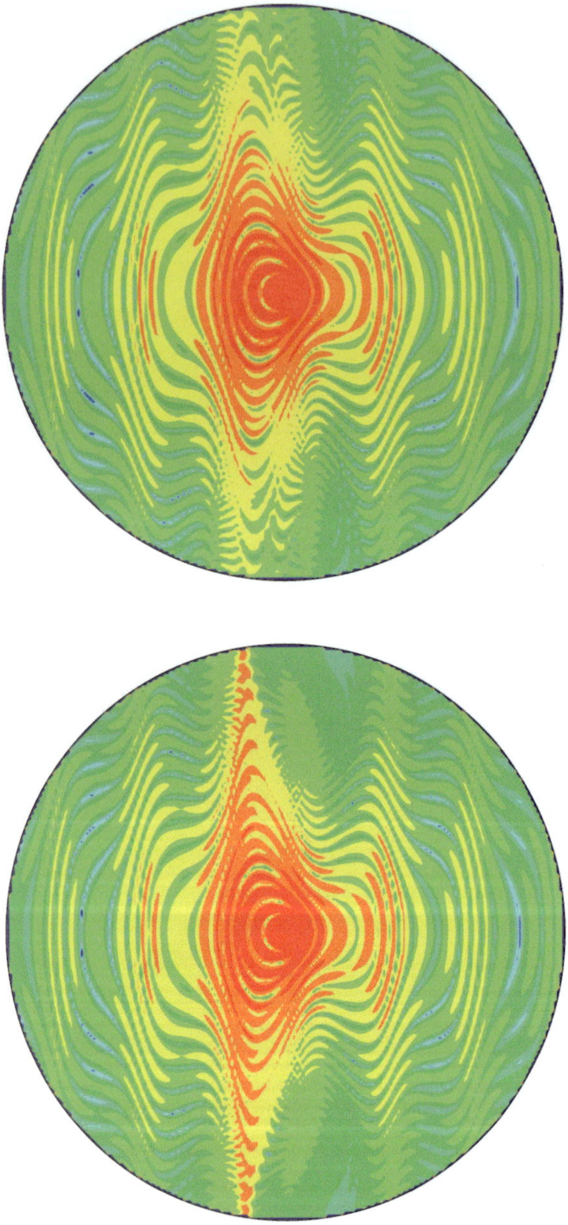

Fig. 2.23 Transmission loss at $z = 30$ m that was obtained for $r < 20$ km by ignoring (top) and accounting for (bottom) azimuthal coupling. The dynamic range is 30 dB, with red corresponding to 55 dB and dark blue corresponding to 115 dB. In the solution that accounts for azimuthal coupling, there is an enhancement of energy in the valley between $x = -3$ km and $x = 0$, where the rays in Fig. 2.22 are horizontally ducted

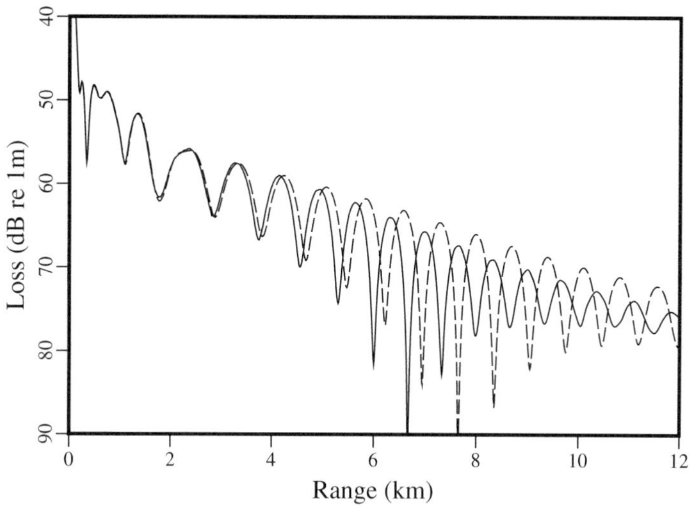

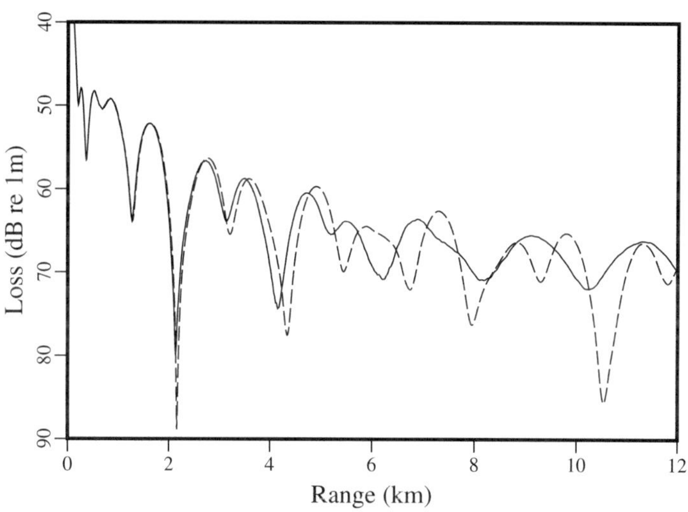

Fig. 2.24 Transmission loss at $z = 30$ m and $\theta = 80^\circ$ (top) and $\theta = 100^\circ$ (bottom) for a problem with a corrugated bottom. The solid curves were obtained by accounting for azimuthal coupling. The dashed curves were obtained by ignoring azimuthal coupling

$$\frac{\partial^2 p}{\partial r^2} + \frac{1}{r^2}\frac{\partial^2 p}{\partial \theta^2} + \rho \frac{\partial}{\partial z}\frac{1}{\rho}\frac{\partial p}{\partial z} + k^2 p = 0. \tag{2.71}$$

We approximate an environment that has variations in both range and azimuth in terms of a set of regions in which there are no horizontal variations. Within each region, the solution may be expressed in terms of the local normal modes. We consider a solution of the form,

$$p(r, z, \theta) = p_m(r, \theta)\zeta_m(z; r, \theta), \tag{2.72}$$

and assume that horizontal variations are sufficiently gradual so that the coupling of energy between modes may be neglected. Substituting into Eq. (2.71) and applying Eq. (1.25), we obtain the horizontal wave equation,

$$\frac{\partial^2 p_m}{\partial r^2} + \frac{1}{r^2}\frac{\partial^2 p_m}{\partial \theta^2} + k_m^2(r, \theta)p_m = 0, \tag{2.73}$$

which may be solved with the parabolic equation techniques that are used to solve Eq. (1.13). For some problems, it is practical to solve Eq. (2.73) out to ranges at which the curvature of the Earth must be taken into account. Such problems may be handled with the global-scale equation [7],

$$\frac{\partial^2 p_m}{\partial r^2} + \frac{1}{R_0^2 \sin^2(r/R_0)}\frac{\partial^2 p_m}{\partial \theta^2} + k_m^2(r, \theta)p_m = 0, \tag{2.74}$$

where R_0 is the radius of the planet. To test the approach, we consider a problem for which $R_0 = 12$ km, the bathymetry,

$$d(x, y) = \left[3 - \sin\left(\frac{2\pi x}{12\ \text{km}}\right)\right] 50\ \text{m}, \tag{2.75}$$

is displayed in Fig. 2.25, a 25 Hz source is 25 m below the surface of the water column, and the sediment parameters are $c = 1700$ m/s, $\rho = 1.5$ g/cm^3, and $\sigma = 0.5$. All three of the propagating modes are included in the solutions appearing in Figs. 2.26 and 2.27 that were obtained with and without azimuthal coupling. There are significant differences in these solutions on the backside. In Fig. 2.28, the

Fig. 2.25 Sinusoidal bathymetry on a planet of radius 12 km, with red corresponding to 200 m and blue corresponding to 100 m

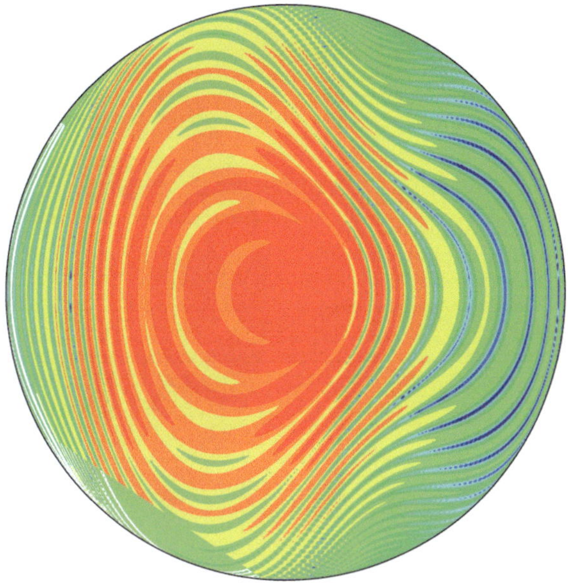

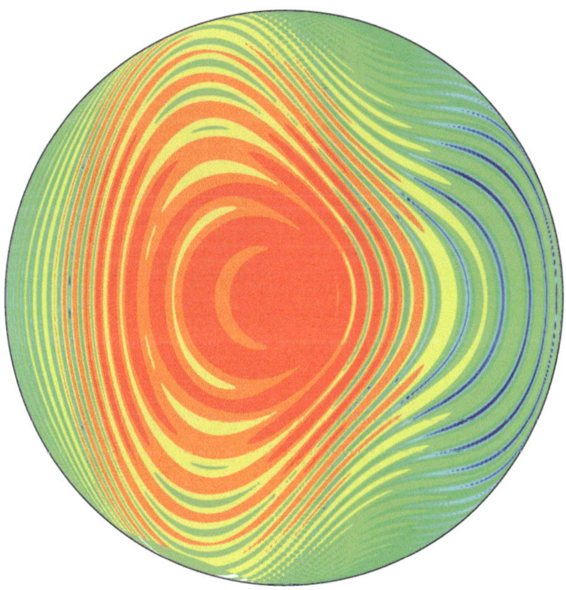

Fig. 2.26 Transmission loss at $z = 30$ m on the front side. Solutions that ignore (top) and account for (bottom) azimuthal coupling. The dynamic range is 40 dB with red corresponding to 60 dB and dark blue corresponding to 100 dB

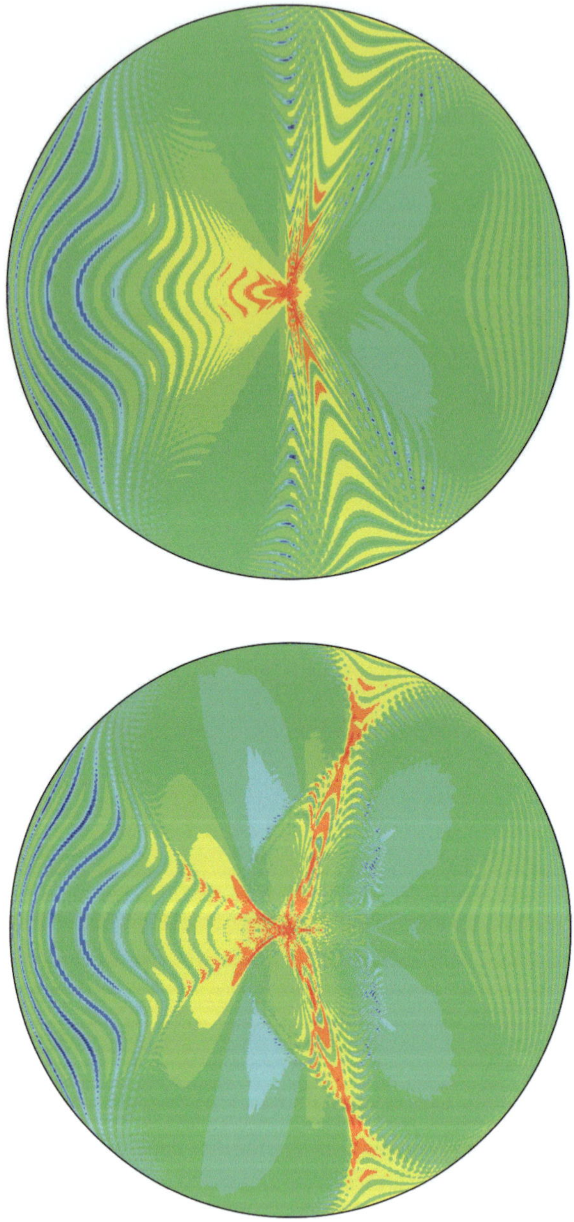

Fig. 2.27 Transmission loss at $z = 30$ m on the backside. Solutions that ignore (top) and account for (bottom) azimuthal coupling. The solution breaks down as the range approaches the antipode, where it is terminated. The dynamic range is 40 dB with red corresponding to 60 dB and dark blue corresponding to 100 dB

solution for the second mode is compared with the ray solution, which has similar features. The global-scale approach and the bathymetry appearing in Fig. 2.29 [34] were used with and without azimuthal coupling to obtain the solutions for the first

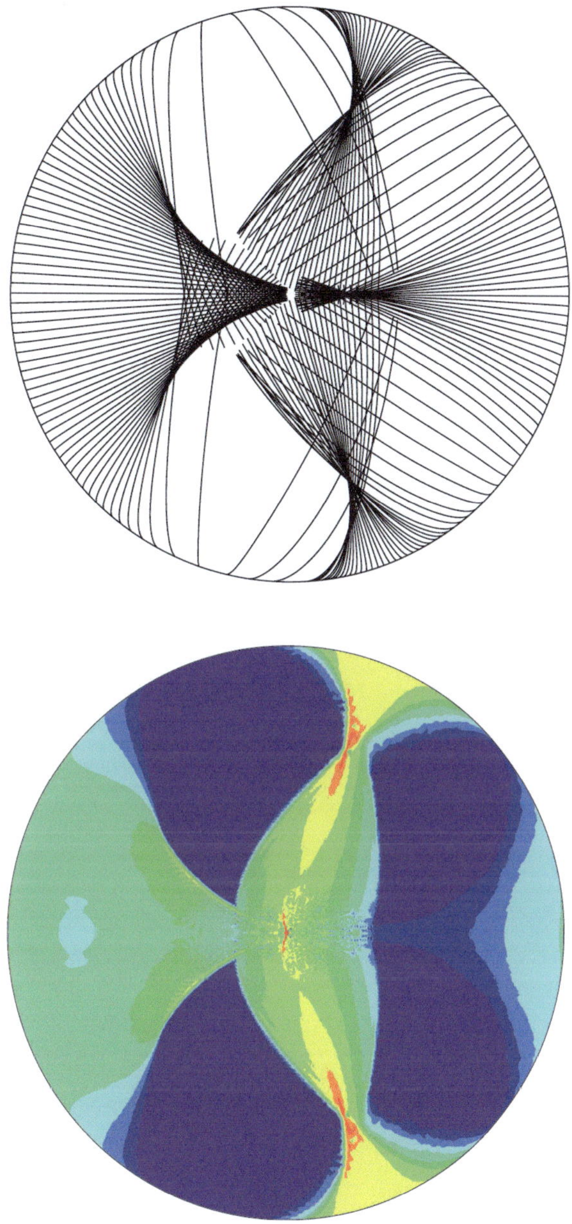

Fig. 2.28 Rays (top) and parabolic equation solution (bottom) for the second mode

mode at 1 Hz that appear in Figs. 2.30 and 2.31 [8]. Interaction with the Hawaiian Islands causes energy to enter the shadow zone and a diffraction pattern to form to the north of the shadow zone.

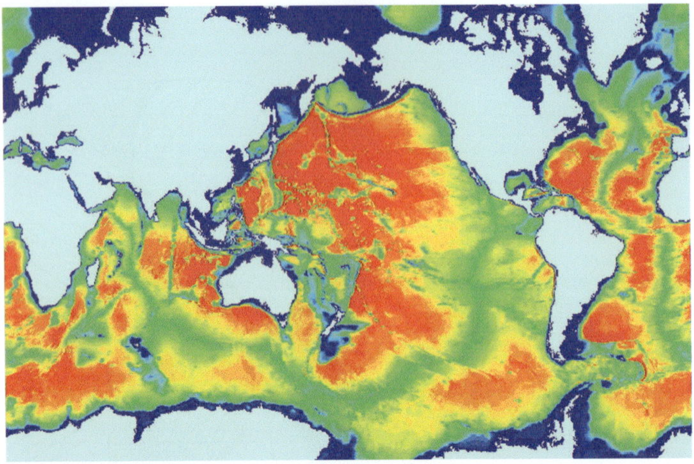

Fig. 2.29 Bathymetry [34] that can be used in global-scale calculations

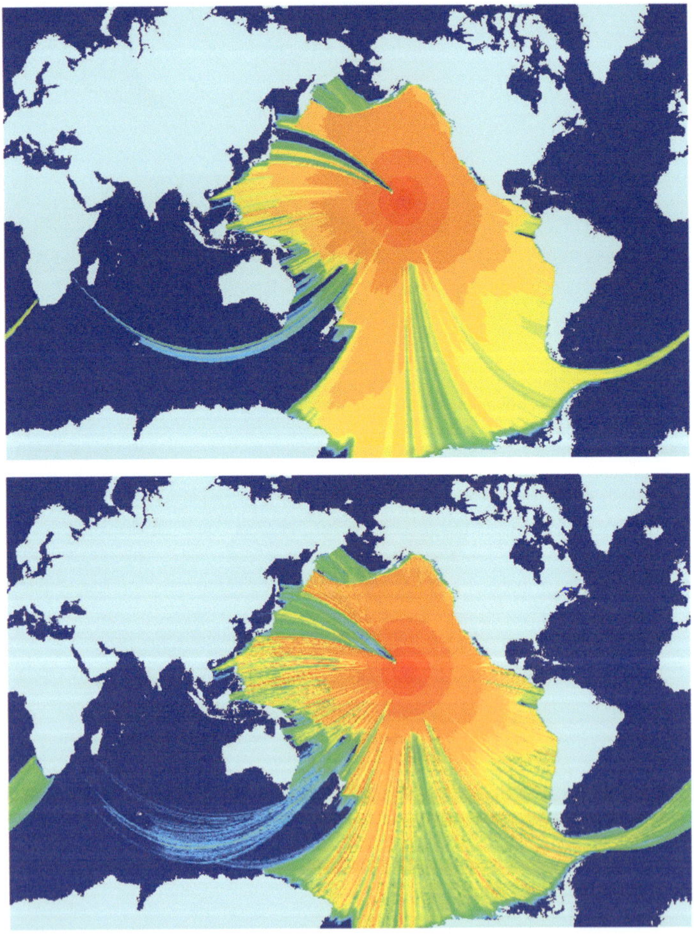

Fig. 2.30 Transmission loss for the first mode at 1 Hz for a source to the southeast of the Hawaiian Islands. There is a deep shadow behind the islands in the solution that neglects azimuthal coupling (top). Energy makes it into the shadow in the solution that accounts for azimuthal coupling (bottom)

Fig. 2.31 Enlargements of the results in Fig. 2.30 that show details of the shadow zone in the solution that neglects azimuthal coupling (top) and the diffraction pattern to the north of the shadow zone in the solution that accounts for azimuthal coupling (bottom)

Appendix: Recursion Formula for Derivatives

In this appendix, we derive a recursion formula for computing the derivatives of the function,

$$f(X) = (1 - \varepsilon X)^2 (1 + X)^{-1/2} \exp\left(ik_0 x_0 (1 + X)^{1/2}\right), \tag{2.76}$$

which appears in Eq. (2.38). Differentiating Eq. (2.76), we obtain

$$\frac{df}{dX} = f(X)h(X), \tag{2.77}$$

$$h(X) = -2\nu(1-\varepsilon X)^{-1} - \frac{1}{2}(1+X)^{-1} + \frac{1}{2}ik_0x_0(1+X)^{-1/2}. \tag{2.78}$$

Differentiating Eq. (2.77), we obtain

$$\frac{d^{n+1}f}{dX^{n+1}} = \sum_{j=0}^{n} \binom{n}{j} \frac{d^{n-j}f}{dX^{n-j}} \frac{d^j h}{dX^j}, \tag{2.79}$$

$$\binom{n}{j} = \frac{n!}{(n-j)!j!}. \tag{2.80}$$

The formula in Eq. (2.79) is recursive in the derivatives in *f*(*X*). Each of the derivatives of *h*(*X*) involves three terms.

References

1. M.D. Collins, "A split-step Padé solution for the parabolic equation method," J. Acoust. Soc. Am. **93**, 1736–1742 (1993).
2. M.D. Collins, "A self-starter for the parabolic equation method," J. Acoust. Soc. Am. **92**, 2069–2074 (1992).
3. M.D. Collins, "An energy-conserving parabolic equation for elastic media," J. Acoust. Soc. Am. **94**, 975–982 (1993).
4. R.N. Baer, "Propagation through a three-dimensional eddy including effects on an array," J. Acoust. Soc. Am. **69**, 70–75 (1981).
5. W.L. Siegmann, G.A. Kriegsmann, and D. Lee, "A wide-angle three-dimensional parabolic wave equation," J. Acoust. Soc. Am. **78**, 659–664 (1985).
6. M.D. Collins and S.A. Chin-Bing, "A three-dimensional parabolic equation model that includes the effects of rough boundaries," J. Acoust. Soc. Am. **87**, 1104–1109 (1990).
7. M.D. Collins, "The adiabatic mode parabolic equation," J. Acoust. Soc. Am. **94**, 2269–2278 (1993).
8. M.D. Collins, B.E. McDonald, K.D. Heaney, and W.A. Kuperman, "Three-dimensional effects in global acoustics," J. Acoust. Soc. Am. **97**, 1567–1575 (1995).
9. A. Bamberger, B. Engquist, L. Halpern, and P. Joly, "Higher order paraxial wave equation approximations in heterogeneous media," SIAM J. Appl. Math. **48**, 129–154 (1988).
10. M.D. Collins, "Applications and time-domain solution of higher-order parabolic equations in underwater acoustics," J. Acoust. Soc. Am. **86**, 1097–1102 (1989).
11. R.J. Cederberg and M.D. Collins, "Application of an improved self-starter to geoacoustic inversion," IEEE J. Ocean. Eng. **22**, 102–109 (1997).
12. M.D. Collins, "Higher-order Padé approximations for accurate and stable elastic parabolic equations with application to interface wave propagation," J. Acoust. Soc. Am. **89**, 1050–1057 (1991).

13. W.H. Press, B.P. Flannery, S.A. Teukolsky, and W.T. Vetterling, *Numerical Recipes* (Cambridge University Press, Cambridge, 1986).
14. F.A. Milinazzo, C.A. Zala, and G.H. Brooke, " Rational square-root approximations for parabolic equation algorithms," J. Acoust. Soc. Am. **101**, 760–766 (1997).
15. A.R. Mitchell and D.F. Griffiths, *The Finite Difference Method in Partial Differential Equations* (Wiley, New York, 1980).
16. R.B. Evans, "A coupled mode solution for acoustic propagation in a waveguide with stepwise depth variations of a penetrable bottom," J. Acoust. Soc. Am. **74**, 188–195 (1983).
17. H. Schmidt, W. Seong, and J.T. Goh, "Spectral super-element approach to range-dependent ocean acoustic modeling," J. Acoust. Soc. Am. **98**, 465–472 (1995).
18. G. Strang, *Linear Algebra and Its Applications* (Academic Press, New York, 1980).
19. M.D. Collins, "Benchmark calculations for higher-order parabolic equations," J. Acoust. Soc. Am. **87**, 1535–1538 (1990).
20. M.D. Collins, "The stabilized self-starter," J. Acoust. Soc. Am. **106**, 1724–1726 (1999).
21. M.D. Collins, "Wave-number sampling at short range," J. Acoust. Soc. Am. **106**, 2535–2539 (1999).
22. F.B. Jensen and C.M. Ferla, "Numerical solutions of range-dependent benchmark problems," J. Acoust. Soc. Am. **87**, 1499–1510 (1990).
23. M.D. Collins and R.B. Evans, "A two-way parabolic equation for acoustic backscattering in the ocean," J. Acoust. Soc. Am. **91**, 1357–1368 (1992).
24. M.B. Porter, F.B. Jensen, and C.M. Ferla, "The problem of energy conservation in one-way models," J. Acoust. Soc. Am. **89**, 1058–1067 (1991).
25. M.D. Collins and E.K. Westwood, "A higher-order energy-conserving parabolic equation for range-dependent ocean depth, sound speed, and density," J. Acoust. Soc. Am. **89**, 1068–1075 (1991).
26. M.D. Collins, "The rotated parabolic equation and sloping interfaces," J. Acoust. Soc. Am. **87**, 1035–1037 (1990).
27. D.A. Outing, W.L. Siegmann, M.D. Collins, and E.K. Westwood, "Generalization of the rotated parabolic equation to variable slopes," J. Acoust. Soc. Am. **120**, 3534–3538 (2006).
28. M.D. Collins, R.A. Coury, and W.L. Siegmann, "Beach acoustics," J. Acoust. Soc. Am. **97**, 2767–2770 (1995).
29. M.D. Collins and D.K. Dacol, "A mapping approach for handling sloping interfaces," J. Acoust. Soc. Am. **107**, 1937–1942 (2000).
30. COMSOL Multiphysics® v. 5.2 (COMSOL AB, Stockholm, Sweden).
31. E.K. Westwood, "Ray methods for flat and sloping shallow-water waveguides," J. Acoust. Soc. Am. **85**, 1885–1894 (1989).
32. D.A. Outing, W.L. Siegmann, and M.D. Collins, "Scholte-to-Rayleigh conversion and other effects in range-dependent elastic media," IEEE J. Ocean. Eng. **32**, 620–625 (2007).
33. W.H. Munk, "Sound channel in an exponentially stratified ocean, with applications to SOFAR," J. Acoust. Soc. Am. **55**, 220–226 (1974).
34. Digital Bathymetric Data Base 5-minute (DBDB5) (U.S. Naval Oceanographic Office, Stennis Space Center, Mississippi).

Chapter 3
Seismology and Seismo-Acoustics

Contents

3.1 Introduction

This chapter covers parabolic equation techniques for elastic waves, which include compressional, shear, boundary, and interface waves in solid materials. Compressional waves correspond to longitudinal motion (parallel to the direction of propagation) and perturbations in volume. Shear waves correspond to transverse motion (orthogonal to the direction of propagation) and perturbations in shape. Boundary and interface waves propagate along boundaries and interfaces and decay exponentially in the transverse direction. Applications of the elastic parabolic equation include seismology and ocean acoustics problems in which the ocean bottom or ice cover supports shear waves.

M. D. Collins, W. L. Siegmann, *Parabolic Wave Equations with Applications*,
https://doi.org/10.1007/978-1-4939-9934-7_3

3.2 The Elastic Wave Equation

In this section, we discuss the derivation of the elastic wave equation and its basic solutions. We consider a medium that is isotropic and range independent. The first step in the derivation is to apply Newton's second law $\mathbf{F} = m\mathbf{a}$ to a small volume element to obtain the following frequency domain equations in two dimensions [1],

$$\frac{\partial \sigma_{xx}}{\partial x} + \frac{\partial \sigma_{xz}}{\partial z} + \rho\omega^2 u = 0, \tag{3.1}$$

$$\frac{\partial \sigma_{xz}}{\partial x} + \frac{\partial \sigma_{zz}}{\partial z} + \rho\omega^2 w = 0, \tag{3.2}$$

where u is the horizontal displacement, w is the vertical displacement, σ_{xx} is the normal stress on a vertical plane, σ_{zz} is the normal stress on a horizontal plane, and σ_{xz} is the tangential stress on a vertical or horizontal plane. The second step in the derivation is to apply the assumption that the medium is isotropic to obtain the following linear relationship between the stresses and distortions of the solid [1],

$$\sigma_{xx} = (\lambda + 2\mu)\frac{\partial u}{\partial x} + \lambda\frac{\partial w}{\partial z}, \tag{3.3}$$

$$\sigma_{zz} = \lambda\frac{\partial u}{\partial x} + (\lambda + 2\mu)\frac{\partial w}{\partial z}, \tag{3.4}$$

$$\sigma_{xz} = \mu\frac{\partial u}{\partial z} + \mu\frac{\partial w}{\partial x}. \tag{3.5}$$

Substituting into Eqs. (3.1) and (3.2), we obtain the elastic wave equation for a heterogeneous medium,

$$(\lambda + 2\mu)\frac{\partial^2 u}{\partial x^2} + \frac{\partial}{\partial z}\left(\mu\frac{\partial u}{\partial z}\right) + \rho\omega^2 u + (\lambda + \mu)\frac{\partial^2 w}{\partial x \partial z} + \frac{\partial \mu}{\partial z}\frac{\partial w}{\partial x} = 0, \tag{3.6}$$

$$\mu\frac{\partial^2 w}{\partial x^2} + \frac{\partial}{\partial z}\left((\lambda + 2\mu)\frac{\partial w}{\partial z}\right) + \rho\omega^2 w + (\lambda + \mu)\frac{\partial^2 u}{\partial x \partial z} + \frac{\partial \lambda}{\partial z}\frac{\partial u}{\partial x} = 0. \tag{3.7}$$

In a homogeneous layer, we differentiate Eqs. (3.6) and (3.7) and combine to obtain the compressional and shear wave equations,

$$\frac{\partial^2 \Delta}{\partial x^2} + \frac{\partial^2 \Delta}{\partial z^2} + \frac{\rho\omega^2}{\lambda + 2\mu}\Delta = 0, \tag{3.8}$$

$$\frac{\partial^2 \Omega_{xz}}{\partial x^2} + \frac{\partial^2 \Omega_{xz}}{\partial z^2} + \frac{\rho \omega^2}{\mu} \Omega_{xz} = 0, \tag{3.9}$$

$$\Delta = \frac{\partial u}{\partial x} + \frac{\partial w}{\partial z}, \tag{3.10}$$

$$\Omega_{xz} = \frac{\partial u}{\partial z} - \frac{\partial w}{\partial x}. \tag{3.11}$$

Compressional waves satisfy Eq. (3.8) and propagate at the compressional speed,

$$c_p = \left(\frac{\lambda + 2\mu}{\rho} \right)^{1/2}. \tag{3.12}$$

Shear waves satisfy Eq. (3.9) and propagate at the shear speed,

$$c_s = \left(\frac{\mu}{\rho} \right)^{1/2}. \tag{3.13}$$

We also define the compressional and shear wave numbers $k_p = \omega / c_p$ and $k_s = \omega / c_s$ and attenuations σ_p and σ_s in decibels per wavelength. Elastic media also support Rayleigh (solid boundary), Scholte (fluid-solid interface), and Stoneley (solid-solid interface) waves, which propagate along boundaries and interfaces. We define the compressional transmission loss $-20\log_{10}|(\lambda + 2\mu)\Delta|$, which is identical to the acoustic transmission loss when $\mu = 0$, the shear transmission loss $-20\log_{10}|\mu \Omega_{xz}|$, and the normal stress transmission loss $-20\log_{10}|\sigma_{zz}|$, which reduces to the compressional transmission loss in a fluid layer and is continuous across horizontal fluid-solid and solid-solid interfaces. Appearing in Fig. 3.1 is an example related to the mode cutoff problem in Fig. 1.1 in which modes couple into shear wave beams in the sediment. Appearing in Fig. 3.2 is an example in which a Rayleigh wave propagates along varying topography.

3.3 The Elastic Parabolic Equation

Generalizing parabolic equation techniques to the elastic case involves issues that do not arise in the fluid case. The vector nature of the elastic wave equation is a complicating factor. The operator in the wave equation does not factor in standard formulations of elasticity [2]. It is necessary to use higher-order approximations to accurately handle waves that propagate at a wide range of speeds [3, 4]. For the elastic case, some of the eigenvalues are located below the real line in the complex plane, which make it necessary to introduce stability constraints in designing rational approximations [5].

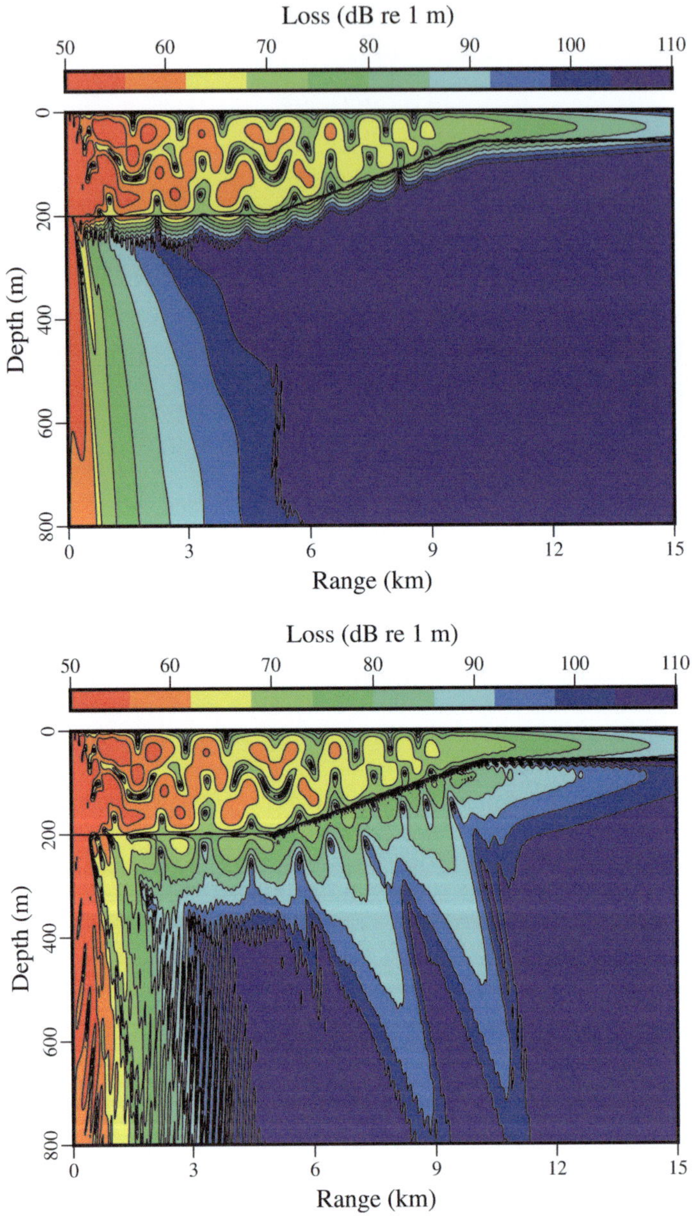

Fig. 3.1 An example with the same bathymetry as the shallow water problem in Fig. 1.1 but with an elastic sediment in which $c_p = 3400$ m/s, $c_s = 1700$ m/s, $\rho = 1.5$ g/cm^3, $\sigma_p = 0.5$, and $\sigma_s = 0.5$. The 25 Hz source is at $z = 25$ m. In the compressional transmission loss (top), there is no sign of coupling of energy into the sediment far from the source. In the normal stress transmission loss (bottom), there is coupling into shear wave beams in the sediment

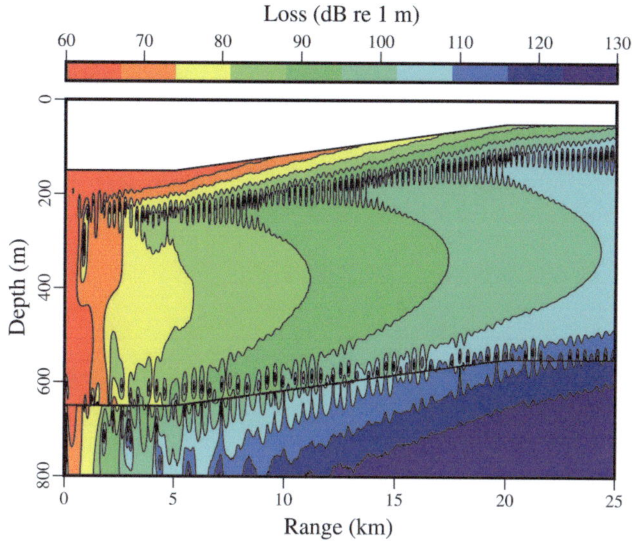

Fig. 3.2 Compressional wave transmission loss for an example in which a Rayleigh wave propagates along a boundary with variable topography. The Rayleigh wave is excited by a 5 Hz source that is located 5 m below the boundary. In the top layer, $c_p = 1700$ m/s, $c_s = 800$ m/s, $\rho = 1.2$ g/cm^3, $\sigma_p = 0.1$, and $\sigma_s = 0.2$. In the lower layer, $c_p = 2400$ m/s, $c_s = 1200$ m/s, $\rho = 1.5$ g/cm^3, $\sigma_p = 0.2$, and $\sigma_s = 0.4$

3.3.1 Factoring the Operator

Parabolic equation techniques are not directly applicable to Eqs. (3.6) and (3.7), which are in the form,

$$\tilde{L}\frac{\partial^2}{\partial x^2}\begin{pmatrix} u \\ w \end{pmatrix} + \tilde{M}\begin{pmatrix} u \\ w \end{pmatrix} = \tilde{N}\frac{\partial}{\partial x}\begin{pmatrix} u \\ w \end{pmatrix}, \tag{3.14}$$

where $\tilde{L}$, $\tilde{M}$, and $\tilde{N}$ are depth operators. Due to the $\partial/\partial x$ term on the right side of Eq. (3.14), the (u, w) formulation of elasticity is not suitable for the parabolic equation method. The operator in the elastic wave equation factors in the (Δ, w) formulation [2], but there are advantages to working in the (u_x, w) formulation [6], where $u_x = \partial u/\partial x$. Differentiating Eq. (3.6) and substituting the new dependent variable, we obtain

$$(\lambda + 2\mu)\frac{\partial^2 u_x}{\partial x^2} + \frac{\partial}{\partial z}\left(\mu\frac{\partial u_x}{\partial z}\right) + \rho\omega^2 u_x + (\lambda + \mu)\frac{\partial^3 w}{\partial x^2 \partial z} + \frac{\partial \mu}{\partial z}\frac{\partial^2 w}{\partial x^2} = 0, \tag{3.15}$$

$$\mu \frac{\partial^2 w}{\partial x^2} + \frac{\partial}{\partial z}\left((\lambda + 2\mu)\frac{\partial w}{\partial z}\right) + \rho\omega^2 w + (\lambda + \mu)\frac{\partial u_x}{\partial z} + \frac{\partial \lambda}{\partial z} u_x = 0. \tag{3.16}$$

These equations are in the form,

$$\left(L\frac{\partial^2}{\partial x^2} + M\right)\mathbf{U} = 0, \tag{3.17}$$

$$\mathbf{U} = \begin{pmatrix} u_x \\ w \end{pmatrix}, \tag{3.18}$$

where L and M are depth operators. Rearranging Eq. (3.17) into a generalization of Eq. (1.46), we obtain

$$\left(\frac{\partial^2}{\partial x^2} + k_0^2(1 + X)\right)\mathbf{U} = 0, \tag{3.19}$$

$$X = k_0^{-2}\left(L^{-1}M - k_0^2\right). \tag{3.20}$$

Factoring the operator in Eq. (3.19) and assuming that outgoing energy dominates incoming energy, we obtain the elastic parabolic equation,

$$\frac{\partial \mathbf{U}}{\partial x} = ik_0(1 + X)^{1/2}\mathbf{U}. \tag{3.21}$$

The solution may be efficiently marched in range using the following generalization of Eq. (2.27),

$$\mathbf{U}(x + \Delta x) = \exp\left(ik_0\Delta x(1 + X)^{1/2}\right)\mathbf{U}(x). \tag{3.22}$$

3.3.2 Boundaries and Coupling with Fluid Layers

The discretization of Eq. (3.17) by Galerkin's method is valid for problems involving continuous variations within solid layers and interfaces between solid layers [6]. Conditions for vanishing stresses may be applied to handle a horizontal boundary. From Eq. (3.4), we obtain the following expression for the normal stress in the (u_x, w) formulation,

$$\sigma_{zz} = \lambda u_x + (\lambda + 2\mu)\frac{\partial w}{\partial z}. \tag{3.23}$$

Since the tangential stress vanishes at all points along a horizontal boundary, its tangential derivative also vanishes [2], and we obtain the following expression for this quantity from Eqs. (3.5) and (3.16):

$$-\frac{\partial \sigma_{xz}}{\partial x} = \frac{\partial}{\partial z}\left(\lambda u_x + (\lambda + 2\mu)\frac{\partial w}{\partial z}\right) + \rho\omega^2 w. \tag{3.24}$$

For problems involving fluid layers, we combine Eqs. (3.6) and (3.7) for the case $\mu = 0$ and obtain

$$\frac{\partial^2}{\partial x^2}(\lambda\Delta) + \rho\frac{\partial}{\partial z}\left(\frac{1}{\rho}\frac{\partial}{\partial z}(\lambda\Delta)\right) + \frac{\rho\omega^2}{\lambda}(\lambda\Delta) = 0. \tag{3.25}$$

This equation corresponds to Eq. (1.14) with $p = -\lambda\Delta$ and $k^2 = \rho\omega^2/\lambda$. Elastic layer A is coupled to fluid layer B by enforcing the conditions,

$$w_A = \frac{1}{\rho_B\omega^2}\frac{\partial p_B}{\partial z}, \tag{3.26}$$

$$\lambda_A(u_x)_A + (\lambda_A + 2\mu_A)\frac{\partial w_A}{\partial z} = -p_B, \tag{3.27}$$

$$\frac{\partial}{\partial z}\left(\lambda_A(u_x)_A + (\lambda_A + 2\mu_A)\frac{\partial w_A}{\partial z}\right) + \rho_A\omega^2 w_A = 0, \tag{3.28}$$

which correspond to continuity of w and σ_{zz} and vanishing of σ_{xz}. The operator in Eq. (3.25) and the interface conditions in Eqs. (3.26)–(3.28) can be incorporated into the operators L and M without changing the form of Eq. (3.17). In fluid layers, the first row of Eq. (3.17) corresponds to Eq. (3.25), and the second row is empty.

3.3.3 Depth Discretization

The depth operators in Eqs. (3.15), (3.16), and (3.25) may be implemented using the finite-difference formulas in Sect. 1.4.1. This approach accounts for continuous variations within layers and discontinuous changes at fluid-fluid and solid-solid interfaces, but fluid-solid interfaces must be handled explicitly in terms of finite-difference formulas for Eqs. (3.26)–(3.28). We use the grid structure illustrated in Fig. 3.3 to facilitate the treatment of a sloping fluid-solid interface, which is approximated in terms of a series of stair steps. The interface is located halfway between the last grid point in the fluid and the first grid point in the solid. The grid

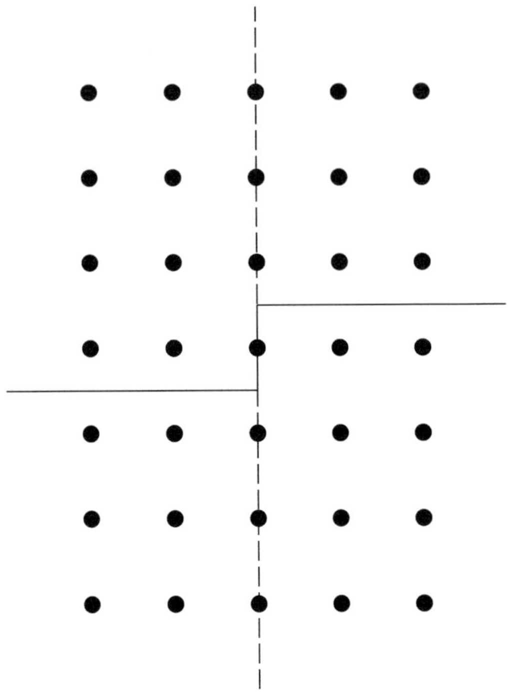

Fig. 3.3 Grid structure for handling a sloping fluid-solid interface, which is approximated in terms of a series of stair steps with rises and runs. The two solid horizontal lines are runs, which are located halfway between grid points. The dashed vertical line is a vertical interface between two range-independent regions that includes a rise. The grid point on the rise is located in the fluid (upper layer) in the incident region (to the left) and is located in the solid (lower layer) in the transmitted region (to the right)

point in the center of the rise of the stair step in Fig. 3.3 is the last grid point in the fluid for the incident field and the first grid point in the solid for the transmitted field. In order to implement the interface conditions, we apply the approach involving artificial grid points that is discussed in Sect. 1.4.3.

Finite-difference formulas for the interface conditions are used to express the dependent variables at the artificial grid points in terms of the dependent variables at the real grid points. The standard three-point finite-difference for the second derivative is not suitable for the grid structure in Fig. 3.3. In order to retain accuracy to second order in h, we approximate the second derivative in Eq. (3.28) in terms of the four-point finite-difference formula [7],

$$\frac{\partial^2 w}{\partial z^2}\left(z-\frac{1}{2}h\right)=\frac{3w(z-h)-7w(z)+5w(z+h)-w(z+2h)}{2h^2}+O\left(h^2\right). \quad (3.29)$$

The other depth operators that appear in the interface conditions may be approximated using the finite-difference formulas in Sect. 1.4.1. There is one artificial quantity at the artificial grid point in the fluid, there are two artificial quantities at the artificial grid point in the solid, and there are three linear equations (finite-difference formulas for the interface conditions) involving these quantities and the dependent variables at real grid points near the interface. At the last grid point in the fluid and the first grid point in the solid, the finite-difference formulas for Eqs. (3.15), (3.16), and (3.25) also involve the artificial quantities. By substituting expressions for the artificial quantities that are obtained using the interface conditions, the entries of the matrices that represent the depth operators are modified near the interface.

Due to the four-point difference formula, nonzero entries are introduced in the modified matrix that are outside the nonzero diagonals in the original matrix as illustrated in Fig. 3.4, but this does not have a significant effect on efficiency. This system of equations may be solved by eliminating entries below the main diagonal during a downward sweep to the black box that is located just below the interface, eliminating entries above the main diagonal during an upward sweep to the black box, solving the 2×2 system in the black box, sweeping downward to obtain the solution below the black box, and sweeping upward to determine the solution above the black box. It is necessary to perform the last two steps in the specified order due to entries to the right in rows near the interface.

3.4 The Self-Starter

For the case of a source in the fluid, the first row of Eq. (3.19) is modified by placing a source term on the right side, but this does not affect the equations in the solid layer. Generalizing Eq. (2.34) for this case, we obtain

$$\mathbf{U}(r_0, z) = \left(\frac{2\pi i}{k_0 r_0}\right)^{1/2} (1+X)^{-1/4} \exp\left(ik_0 r_0 (1+X)^{1/2}\right) \begin{pmatrix} \delta(z-z_0) \\ 0 \end{pmatrix}. \tag{3.30}$$

Appearing in Fig. 3.5 is an example involving a Scholte wave propagating along a fluid-solid interface at the bottom of the ocean. The self-starter properly excites the Scholte wave and a mode that correspond to the peaks in the wave number spectrum and form an interference pattern near the interface.

3.4.1 Compressional and Shear Potentials

For the case of a source in the solid, we assume that the medium is homogeneous in a depth interval that contains the source and use the representation,

$$u = \frac{\partial \phi}{\partial x} + \frac{\partial \psi}{\partial z}, \tag{3.31}$$

$$w = \frac{\partial \phi}{\partial z} - \frac{\partial \psi}{\partial x}, \tag{3.32}$$

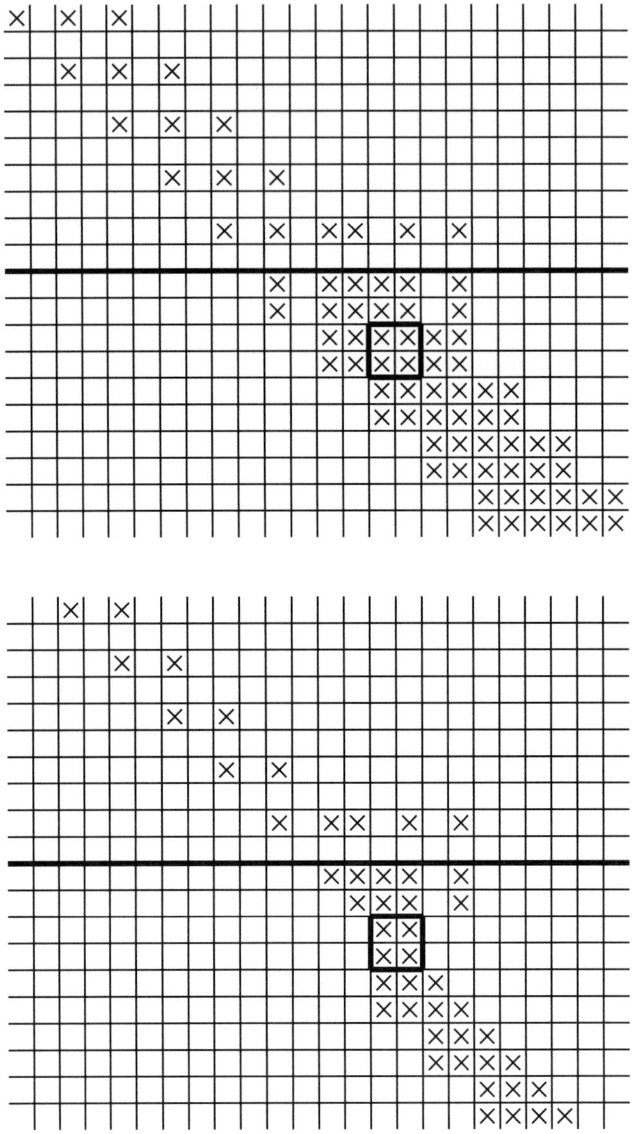

Fig. 3.4 The structure of matrices that correspond to depth operators in the seismo-acoustic parabolic equation solution before (top) and after (bottom) the elimination steps. The bold horizontal line corresponds to the fluid-solid interface. In the fluid layer (above the interface), the even rows are empty and the odd rows contain three nonzero entries away from the interface. In the solid layer (below the interface), there are six nonzero entries per row away from the interface. Rows near the interface contain additional nonzero entries above the main diagonal that correspond to the four-point difference formula. The bold box corresponds to the 2×2 system that remains after the elimination steps

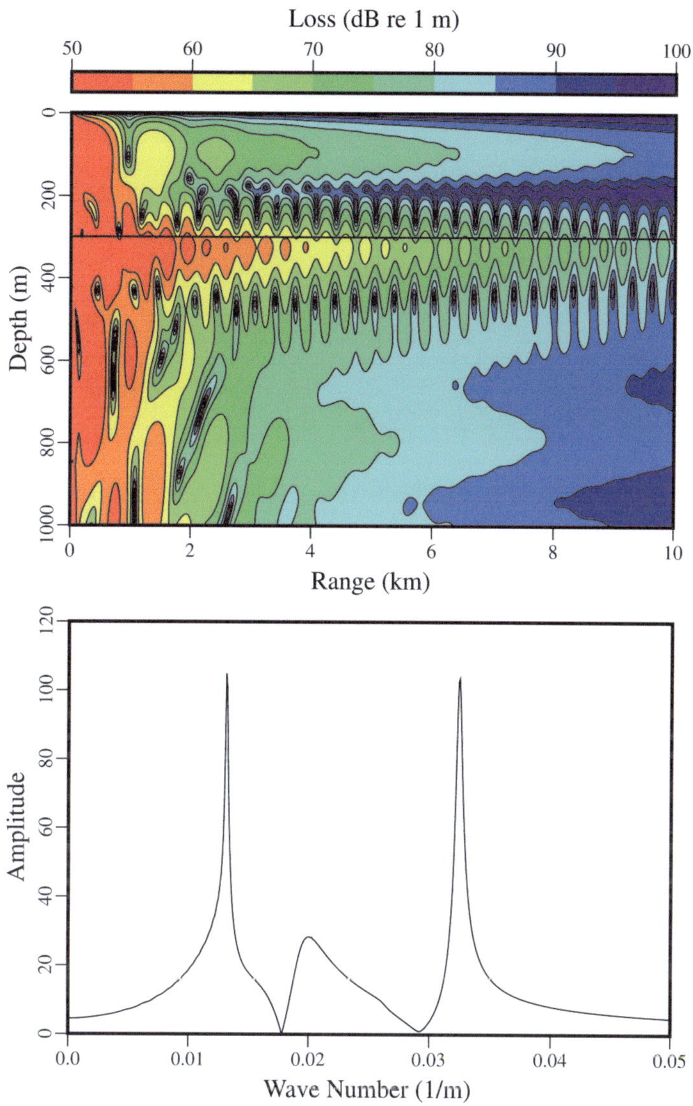

Fig. 3.5 Compressional wave transmission loss that was generated using the self-starter and the elastic parabolic equation (top) and the wave number spectrum at $z = 285$ m (bottom) for an example involving a 5 Hz source at $z = 285$ m in a 300 m deep ocean. In the sediment, $c_p = 2400$ m/s, $c_s = 1200$ m/s, $\rho = 1.5$ g/cm^3, $\sigma_p = 0.5$, and $\sigma_s = 0.5$. In the spectrum, the left peak corresponds to a mode and the right peak corresponds to a Scholte wave

$$\frac{\partial^2 \phi}{\partial x^2} + \frac{\partial^2 \phi}{\partial z^2} + k_p^2 \phi = 0, \tag{3.33}$$

$$\frac{\partial^2 \psi}{\partial x^2} + \frac{\partial^2 \psi}{\partial z^2} + k_s^2 \psi = 0, \tag{3.34}$$

where ϕ and ψ are the compressional and shear potentials [1]. From Eqs. (3.31)–(3.34), we obtain

$$(\lambda + 2\mu) u_x + (\lambda + 2\mu) \frac{\partial w}{\partial z} = (\lambda + 2\mu) \left(\frac{\partial^2 \phi}{\partial x^2} + \frac{\partial^2 \phi}{\partial z^2} \right) = -\rho \omega^2 \phi, \tag{3.35}$$

$$\mu \left(\frac{\partial u_x}{\partial z} - \frac{\partial^2 w}{\partial x^2} \right) = \mu \left(\frac{\partial^2}{\partial x^2} + \frac{\partial^2}{\partial z^2} \right) \frac{\partial \psi}{\partial x} = -\rho \omega^2 \frac{\partial \psi}{\partial x}. \tag{3.36}$$

From Eqs. (3.16) and (3.36), we obtain

$$(\lambda + 2\mu) \frac{\partial u_x}{\partial z} + (\lambda + 2\mu) \frac{\partial^2 w}{\partial z^2} + \rho \omega^2 w = -\rho \omega^2 \frac{\partial \psi}{\partial x}. \tag{3.37}$$

3.4.2 Compressional Source in a Solid Layer

Placing a term corresponding to a compressional source at $z = z_0$ on the right side of Eq. (3.33), we obtain

$$\frac{\partial^2 \phi}{\partial x^2} + \frac{\partial^2 \phi}{\partial z^2} + k_p^2 \phi = -2i\delta(x)\delta(z - z_0). \tag{3.38}$$

Integrating Eq. (3.38) about an arbitrarily small interval about $x = 0$, we obtain

$$\lim_{x \to 0^+} \frac{\partial \phi}{\partial x} = -i\delta(z - z_0). \tag{3.39}$$

From Eqs. (3.21), (3.35), (3.37), and (3.39), we obtain

$$\lim_{x\to 0^+}\begin{pmatrix} \lambda+2\mu & (\lambda+2\mu)\dfrac{\partial}{\partial z} \\ (\lambda+2\mu)\dfrac{\partial}{\partial z} & (\lambda+2\mu)\dfrac{\partial^2}{\partial z^2}+\rho\omega^2 \end{pmatrix}\left(L^{-1}M\right)^{1/2}\mathbf{U}(x,z)$$
$$=\begin{pmatrix} \rho\omega^2\delta(z-z_0) \\ 0 \end{pmatrix}. \tag{3.40}$$

Dividing both rows of Eq. (3.40) by $\lambda + 2\mu$, we obtain

$$\lim_{x\to 0^+}\begin{pmatrix} 1 & \dfrac{\partial}{\partial z} \\ \dfrac{\partial}{\partial z} & \dfrac{\partial^2}{\partial z^2}+k_p^2 \end{pmatrix}\left(L^{-1}M\right)^{1/2}\mathbf{U}(x,z)=\begin{pmatrix} k_p^2\delta(z-z_0) \\ 0 \end{pmatrix}. \tag{3.41}$$

Subtracting the depth derivative of the top row from the bottom row and then dividing the bottom row by k_p^2, we obtain

$$\lim_{x\to 0^+}\begin{pmatrix} 1 & \dfrac{\partial}{\partial z} \\ 0 & 1 \end{pmatrix}\left(L^{-1}M\right)^{1/2}\mathbf{U}(x,z)=\begin{pmatrix} k_p^2\delta(z-z_0) \\ -\delta'(z-z_0) \end{pmatrix}. \tag{3.42}$$

Subtracting the depth derivative of the bottom row from the top row, we obtain

$$\lim_{x\to 0^+}\left(L^{-1}M\right)^{1/2}\mathbf{U}(x,z)=\begin{pmatrix} k_p^2\delta(z-z_0)+\delta''(z-z_0) \\ -\delta'(z-z_0) \end{pmatrix}. \tag{3.43}$$

Applying Eq. (3.22) to march the field out a short distance in range from the source singularity, we obtain the self-starter,

$$\mathbf{U}(x_0,z)=\left(L^{-1}M\right)^{-1/2}\exp\left(ix_0\left(L^{-1}M\right)^{1/2}\right)$$
$$\times\begin{pmatrix} k_p^2\delta(z-z_0)+\delta''(z-z_0) \\ -\delta'(z-z_0) \end{pmatrix}, \tag{3.44}$$

which can be implemented using the approach discussed in Sect. 2.5. For the case of a point source in cylindrical geometry, we obtain

$$\mathbf{U}(r_0,z)=r_0^{-1/2}\left(L^{-1}M\right)^{-1/4}\exp\left(ir_0\left(L^{-1}M\right)^{1/2}\right)\\ \times\begin{pmatrix} k_p^2\delta(z-z_0)+\delta''(z-z_0)\\ -\delta'(z-z_0)\end{pmatrix}. \tag{3.45}$$

The delta function and its derivatives appearing on the right side of Eq. (3.45) may be handled using the approaches described in Sect. 1.4.2. Appearing in Fig. 3.6 is an example involving an array of 25 Hz compressional sources that are phased to excite a beam that propagates upward toward the surface at an angle of -35° relative to the horizontal. The self-starter properly excites the compressional wave beam but does not excite shear waves. A shear wave beam is excited when the compressional wave beam reflects from the boundary.

3.4.3 Shear Source in a Solid Layer

Placing a term corresponding to a shear source at $z = z_0$ on the right side of Eq. (3.34), we obtain

$$\frac{\partial^2\psi}{\partial x^2}+\frac{\partial^2\psi}{\partial z^2}+k_s^2\psi=2\delta(x)\delta(z-z_0). \tag{3.46}$$

Integrating Eq. (3.46) about an arbitrarily small interval about $x = 0$, we obtain

$$\lim_{x\to 0^+}\frac{\partial\psi}{\partial x}=\delta(z-z_0). \tag{3.47}$$

From Eqs. (3.35), (3.37), and (3.47), we obtain

$$\lim_{x\to 0^+}\begin{pmatrix} \lambda+2\mu & (\lambda+2\mu)\frac{\partial}{\partial z}\\ (\lambda+2\mu)\frac{\partial}{\partial z} & (\lambda+2\mu)\frac{\partial^2}{\partial z^2}+\rho\omega^2\end{pmatrix}\mathbf{U}(x,z)=\begin{pmatrix}0\\ \rho\omega^2\delta(z-z_0)\end{pmatrix}. \tag{3.48}$$

Subtracting the depth derivative of the top row from the bottom row and then dividing the bottom row by $\rho\omega^2$ and the top row by $\lambda + 2\mu$, we obtain

$$\lim_{x\to 0^+}\begin{pmatrix} 1 & \frac{\partial}{\partial z}\\ 0 & 1\end{pmatrix}\mathbf{U}(x,z)=\begin{pmatrix}0\\ \delta(z-z_0)\end{pmatrix}. \tag{3.49}$$

Subtracting the depth derivative of the bottom row from the top row, we obtain

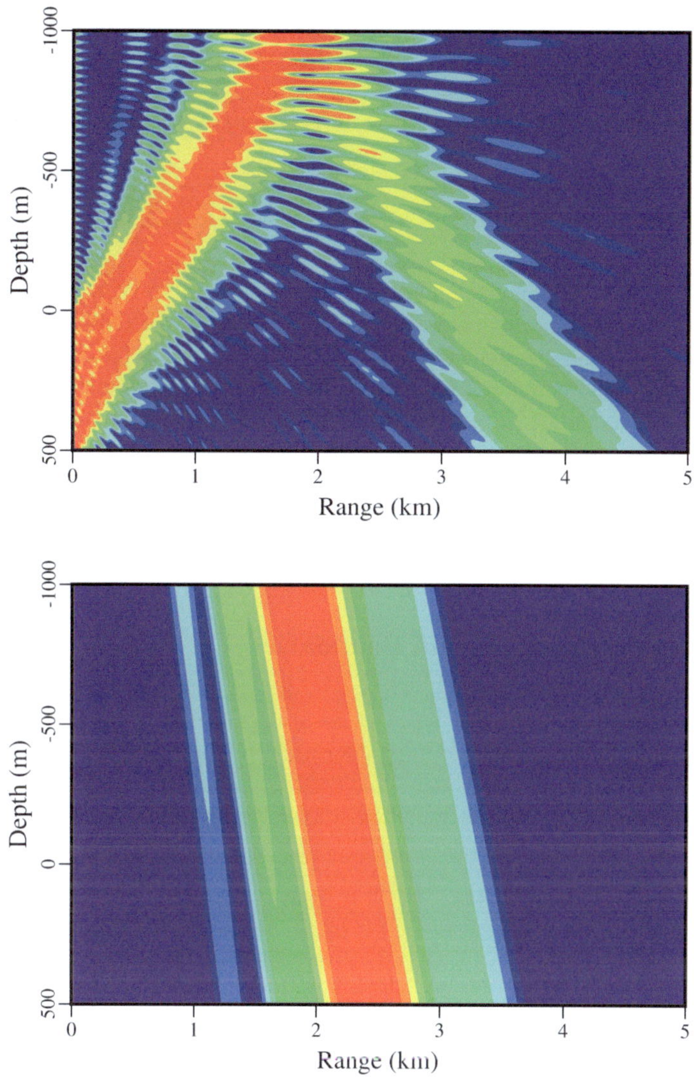

Fig. 3.6 Compressional (top) and shear (bottom) transmission loss (20 dB dynamic range for each plot) for an example involving an array of twenty compressional sources at 25 Hz in a lossless half space in which $c_p = 1500$ m/s and $c_s = 700$ m/s. The jth source is at a depth of $z_j = (j-1)30$ m (half wavelength spacing) and has the phase factor $\exp(-ik_p z_j \sin 35°)$

$$\lim_{x \to 0^+} \mathbf{U}(x, z) = \begin{pmatrix} -\delta'(z - z_0) \\ \delta(z - z_0) \end{pmatrix}. \tag{3.50}$$

Applying Eq. (3.22) to march the field out a short distance in range from the source singularity, we obtain the self-starter,

$$\mathbf{U}(x_0, z) = \exp\left(ix_0\left(L^{-1}M\right)^{1/2}\right)\begin{pmatrix} -\delta'(z-z_0) \\ \delta(z-z_0) \end{pmatrix}. \tag{3.51}$$

For the case of a point source in cylindrical geometry, we obtain

$$\mathbf{U}(r_0, z) = r_0^{-1/2}\left(L^{-1}M\right)^{1/4}\exp\left(ir_0\left(L^{-1}M\right)^{1/2}\right)\begin{pmatrix} -\delta'(z-z_0) \\ \delta(z-z_0) \end{pmatrix}. \tag{3.52}$$

Appearing in Fig. 3.7 is an example involving an array of 25 Hz shear sources that are phased to excite a beam that propagates upward toward the surface at an angle of -35° relative to the horizontal. The self-starter properly excites the shear wave beam but does not excite compressional waves. The main beam is incident beyond the critical angle for coupling from shear to compressional waves. In the compressional transmission loss, there is only a small evanescent feature near the boundary at the point where the main shear wave beam is incident, but there are beams associated with some of the sidelobes of the shear wave beam that are incident at steeper angles.

3.5 Sloping Interfaces and Boundaries

Several mathematically distinct types of range dependence occur in environments with fluid and solid layers and sloping interfaces and boundaries. We first consider an environment that does not contain fluid layers. At a vertical interface between two range-independent regions, we apply the conversion formulas,

$$\begin{pmatrix} \sigma_{xx} \\ w \end{pmatrix} = R\mathbf{U}, \tag{3.53}$$

$$\begin{pmatrix} u \\ -\sigma_{xz} \end{pmatrix} = -iST^{-1/2}\mathbf{U}, \tag{3.54}$$

$$R = \begin{pmatrix} \lambda + 2\mu & \lambda\dfrac{\partial}{\partial z} \\ 0 & 1 \end{pmatrix}, \tag{3.55}$$

$$S = \begin{pmatrix} 1 & 0 \\ \lambda\dfrac{\partial}{\partial z} + \dfrac{\partial\lambda}{\partial z} & \dfrac{\partial}{\partial z}(\lambda + 2\mu)\dfrac{\partial}{\partial z} + \rho\omega^2 \end{pmatrix}, \tag{3.56}$$

where $T = L^{-1}M$. The following conditions correspond to conservation of the displacements and stresses:

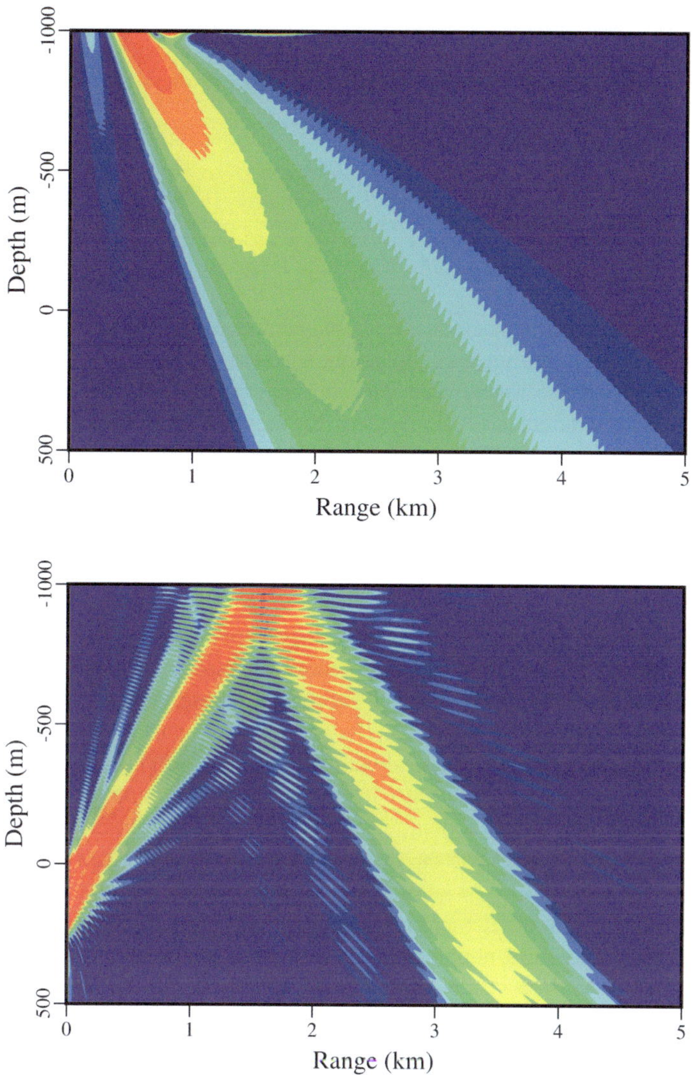

Fig. 3.7 Compressional (top) and shear (bottom) transmission loss (20 dB dynamic range for each plot) for an example involving an array of twenty shear sources at 25 Hz in a lossless half space in which $c_p = 1500$ m/s and $c_s = 700$ m/s. The *j*th source is at a depth of $z_j = (j-1)14$ m (half wavelength spacing) and has the phase factor $\exp(-ik_s z_j \sin 35^\circ)$

$$R_B \mathbf{U}_t = R_A(\mathbf{U}_i + \mathbf{U}_r), \tag{3.57}$$

$$S_B T_B^{-1/2} \mathbf{U}_t = S_A T_A^{-1/2} (\mathbf{U}_i - \mathbf{U}_r). \tag{3.58}$$

The negative sign in Eq. (3.58) accounts for the fact that the reflected field is incoming.

Inverting the operators on the left sides of Eqs. (3.57) and (3.58) and summing, we obtain

$$\begin{aligned}\mathbf{U}_t =&\frac{1}{2}\left(R_B^{-1}R_A + T_B^{1/2}S_B^{-1}S_AT_A^{-1/2}\right)\mathbf{U}_i\\ &+\frac{1}{2}\left(R_B^{-1}R_A - T_B^{1/2}S_B^{-1}S_AT_A^{-1/2}\right)\mathbf{U}_r.\end{aligned} \tag{3.59}$$

Based on the assumption of gradual range dependence, the operator that acts on the reflected field is a difference of operators that are perturbations of the identity operator. Based on the assumption that outgoing energy dominates backscattered energy, the reflected field is small relative to the incident field. Neglecting the term in Eq. (3.59) that is small to second order, we obtain the single-scattering approximation,

$$\mathbf{U}_t = \frac{1}{2}R_B^{-1}R_A\mathbf{U}_i + \frac{1}{2}T_B^{1/2}S_B^{-1}S_AT_A^{-1/2}\mathbf{U}_i. \tag{3.60}$$

Since the parabolic equation has only one range derivative, it is not possible for an outgoing solution to conserve all four quantities (displacements and stresses) across a vertical interface. The single-scattering correction conserves these quantities in an average sense; the first term in Eq. (3.60) corresponds to conservation of σ_{xx} and w; the second term corresponds to conservation of u and σ_{xz}. In a fluid, the first row of Eq. (3.60) corresponds to Eq. (2.49). The single-scattering approximation may be used to account for variable topography by introducing an artificial layer with low wave speeds and density above the boundary, which is then treated as an interface.

We illustrate the accuracy of the single-scattering approximation by making comparisons with a finite-element model [8] for problems involving sloping solid-solid interfaces and solid boundaries [9]. For these examples, a 25 Hz source is located in a 300 m deep water column in which $c = 1500$ m/s, and the basement parameters are $c_p = 3400$ m/s, $c_s = 1700$ m/s, $\rho = 2.5$ g/cm^3, $\sigma_p = 0.2$, and $\sigma_s = 0.4$. For the cases appearing in Figs. 3.8 and 3.9, there are sloping solid-solid interfaces between the basement and a sediment layer in which $c_p = 2400$ m/s, $c_s = 1200$ m/s, $\rho = 1.5$ g/cm^3, $\sigma_p = 0.1$, and $\sigma_s = 0.2$. For the upslope case, the thickness of the sediment is 120 m for $r < 3$ km, 20 m for $r > 7$ km, and linearly decreasing between these values for 3 km $< r <$ 7 km. For the downslope case, the thickness of the sediment is 20 m for $r < 3$ km, 120 m for $r > 7$ km, and linearly increasing between these values for 3 km $< r <$ 7 km. For the cases appearing in Figs. 3.10 and 3.11, there are sloping solid boundaries at the surface of an ice layer in which $c_p = 3500$ m/s, $c_s = 1750$ m/s, $\rho = 0.9$ g/cm^3, $\sigma_p = 0.1$, and $\sigma_s = 0.2$. In an artificial layer above the ice, we take $c_p = 200$ m/s, $c_s = 100$ m/s, $\rho = 0.01$ g/cm^3, $\sigma_p = 10$, and $\sigma_s = 10$. For the upslope case, the thickness of the ice is 20 m for $r < 3$ km, 120 m for $r > 7$ km, and linearly increasing between these values for 3 km $< r <$ 7 km.

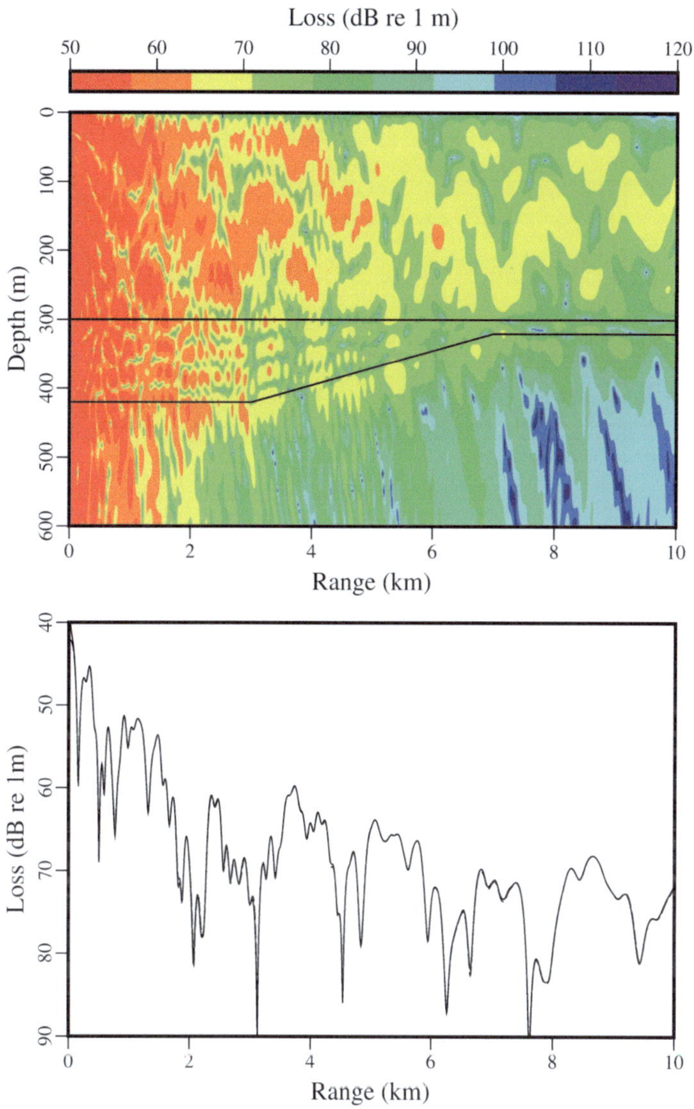

Fig. 3.8 Normal stress transmission loss for a problem involving a 25 Hz source at $z = 250$ m in the water column and an upward sloping solid-solid interface between a sediment layer and a basement. The curves in the lower half of the figure correspond to a receiver at $z = 75$ m. The single-scattering solution (solid curve) is in agreement with the finite-element solution (dashed curve)

For the downslope case, the thickness of the ice is 120 m for $r < 3$ km, 20 m for $r > 7$ km, and linearly decreasing between these values for $3 \text{ km} < r < 7$ km.

Various approaches for handling sloping fluid-solid interfaces have been proposed [10–12]. An approach that provides an appealing combination of accuracy,

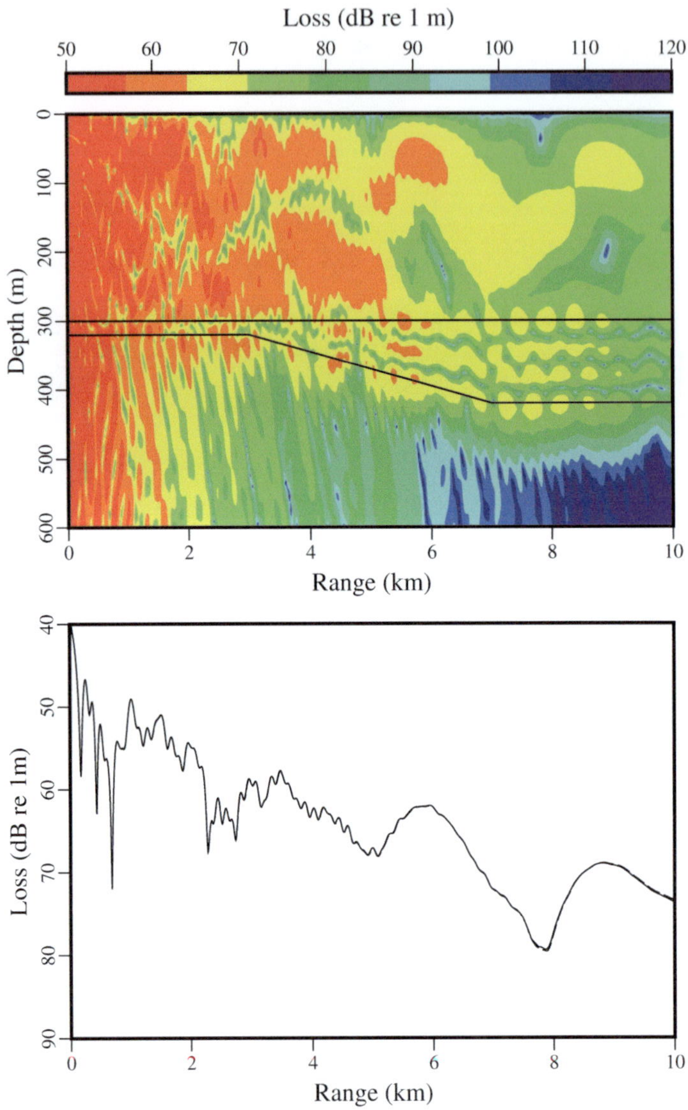

Fig. 3.9 Normal stress transmission loss for a problem involving a 25 Hz source at $z = 250$ m in the water column and a downward sloping solid-solid interface between a sediment layer and a basement. The curves in the lower half of the figure correspond to a receiver at $z = 75$ m. The single-scattering solution (solid curve) is in agreement with the finite-element solution (dashed curve)

simplicity, and efficiency is an approximate generalization of the energy-conservation correction for the acoustic case that accounts for the condition $\sigma_{xz} = 0$ on the rises of the stair steps that approximate a sloping interface. For the acoustic case, we set $\mu = 0$ in Eq. (3.6) and obtain

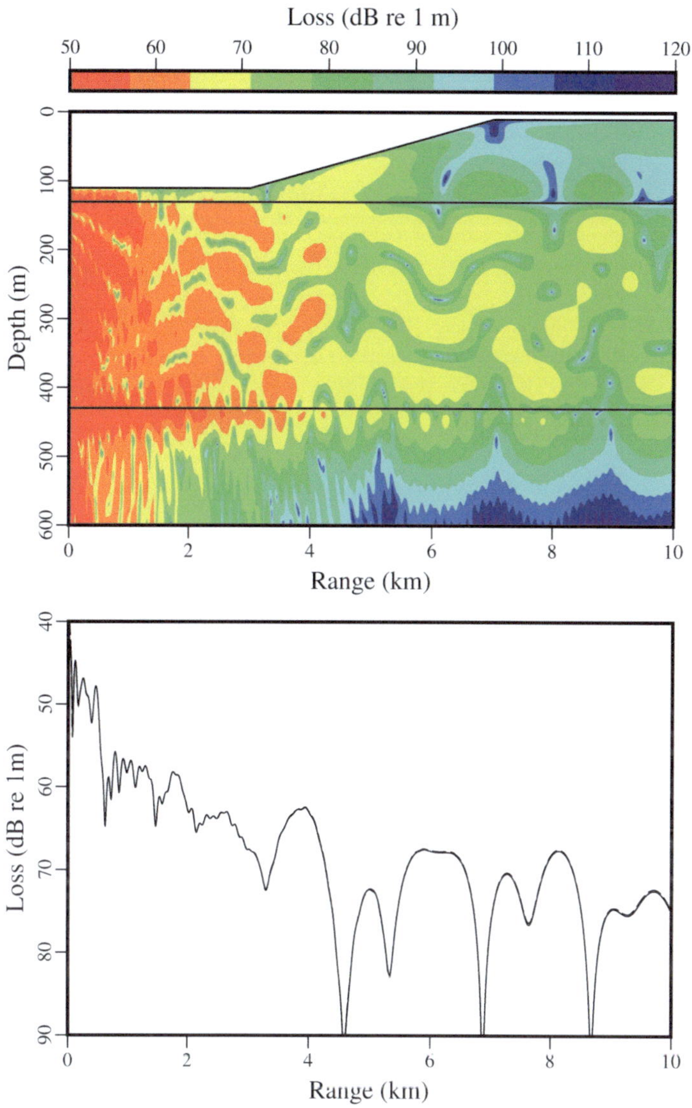

Fig. 3.10 Normal stress transmission loss for a problem involving a source at $z = 410$ m in the water column, which lies between ice with an upward sloping boundary and an elastic basement. The curves in the lower half of the figure correspond to a receiver at $z = 310$ m. The single-scattering solution (solid curve) is in agreement with the finite-element solution (dashed curve)

$$u = -\frac{\lambda}{\rho\omega^2}\frac{\partial}{\partial x}\left(\frac{\partial w}{\partial x} + \frac{\partial w}{\partial z}\right) = \frac{1}{\rho\omega^2}\frac{\partial p}{\partial x}. \tag{3.61}$$

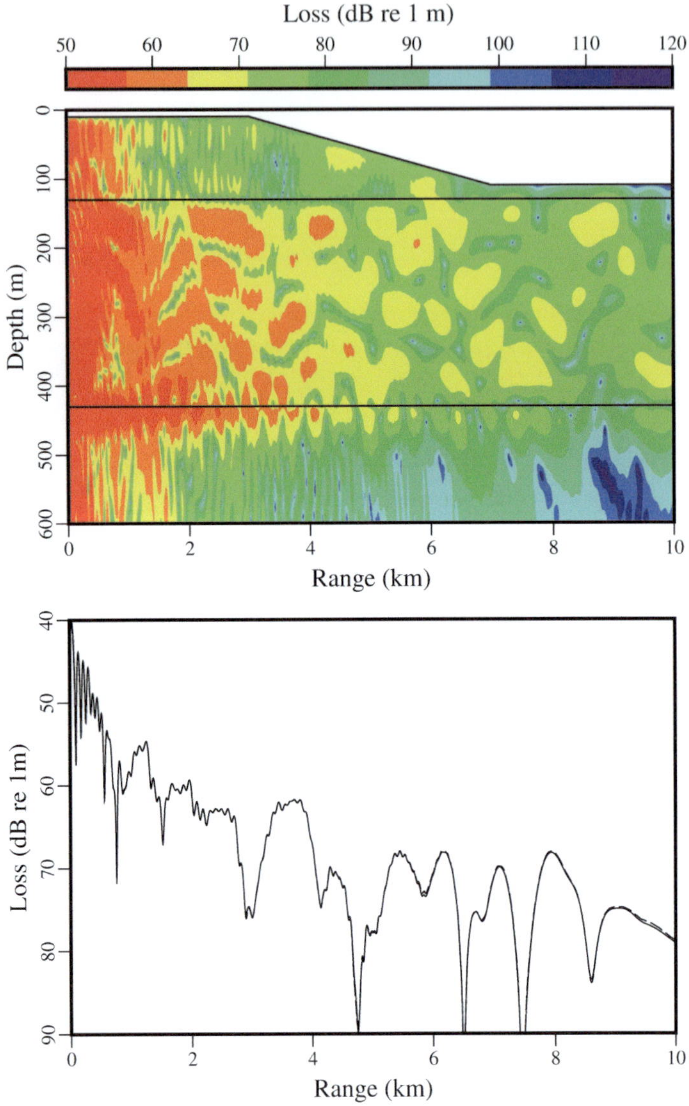

Fig. 3.11 Normal stress transmission loss for a problem involving a source at $z = 410$ m in the water column, which lies between ice with a downward sloping boundary and an elastic basement. The curves in the lower half of the figure correspond to a receiver at $z = 310$ m. The single-scattering solution (solid curve) is in agreement with the finite-element solution (dashed curve)

For the case of a single mode, we use Eq. (3.61) to change variables in the quantity that is conserved in Eq. (2.55) and obtain

$$\rho^{-1/2}k_0^{1/2}(1+X)^{1/4}p = -i\omega^2 k_m^{-1/2}\rho^{1/2}u. \tag{3.62}$$

It follows from Eq. (3.54) that the energy-conservation condition,

$$\rho_B^{1/2}S_B T_B^{-1/2}\mathbf{U}_t = \rho_A^{1/2}S_A T_A^{-1/2}\mathbf{U}_i, \tag{3.63}$$

conserves $\rho^{1/2}u$ and satisfies $\sigma_{xz} = 0$ on the rises. It would be possible to account for the factor of $k_m^{-1/2}$ in Eq. (3.62) by replacing $T^{-1/2}$ with $T^{-3/4}$ in Eq. (3.63), but the condition for $\sigma_{xz} = 0$ would no longer hold with that modification.

We illustrate the accuracy of the energy-conservation condition for problems involving sloping fluid-solid interfaces [12]. For these examples, a 25 Hz source is located in a water column in which $c = 1500$ m/s. For the cases appearing in Figs. 3.12 and 3.13, there is ice cover with variable thickness (but a horizontal surface) and a basement in which $c_p = 3400$ m/s, $c_s = 1700$ m/s, $\rho = 2.5$ g/cm^3, $\sigma_p = 0.2$, and $\sigma_s = 0.4$. The water column is 300 m deep at the source, which is 280 m below the ice. For the downslope case, the thickness of the ice is 20 m for $r < 3$ km, 100 m for $r > 7$ km, and linearly increasing between these values for 3 km $< r <$ 7 km. For the upslope case, the thickness of the ice is 100 m for $r < 3$ km, 20 m for $r > 7$ km, and linearly decreasing between these values for 3 km $< r <$ 7 km. For the cases appearing in Figs. 3.14 and 3.15, there are sloping interfaces between the water column and a basement in which $c_p = 1700$ m/s, $c_s = 800$ m/s, $\rho = 1.2$ g/cm^3, $\sigma_p = 0.2$, and $\sigma_s = 0.4$. For the downslope case, the ocean depth is 400 m for $r < 3$ km, 600 m for $r > 7$ km, and linearly increasing for 3 km $< r <$ 7 km. For the upslope case, the ocean depth is 400 m for $r < 3$ km, 200 m for $r > 7$ km, and linearly decreasing for 3 km $< r <$ 7 km. For problems involving different wave speeds and density, the accuracy of the energy-conservation condition is comparable to the accuracy of the approaches that were tested in Ref. [12]. These approaches may break down when there is a large density contrast at a sloping fluid-solid interface, but this issue can be avoided by introducing a thin layer of sediment with moderate density below the fluid [12].

3.6 Anisotropic Elastic Waves

The wave speeds depend on the direction of propagation in an anisotropic medium. We consider the case of a transverse isotropic medium [13–15], in which Eq. (3.5) remains the same but Eqs. (3.3) and (3.4) generalize to

$$\sigma_{xx} = (\lambda + 2\mu + 2\nu)\frac{\partial u}{\partial x} + (\lambda + 4\kappa)\frac{\partial w}{\partial z}, \tag{3.64}$$

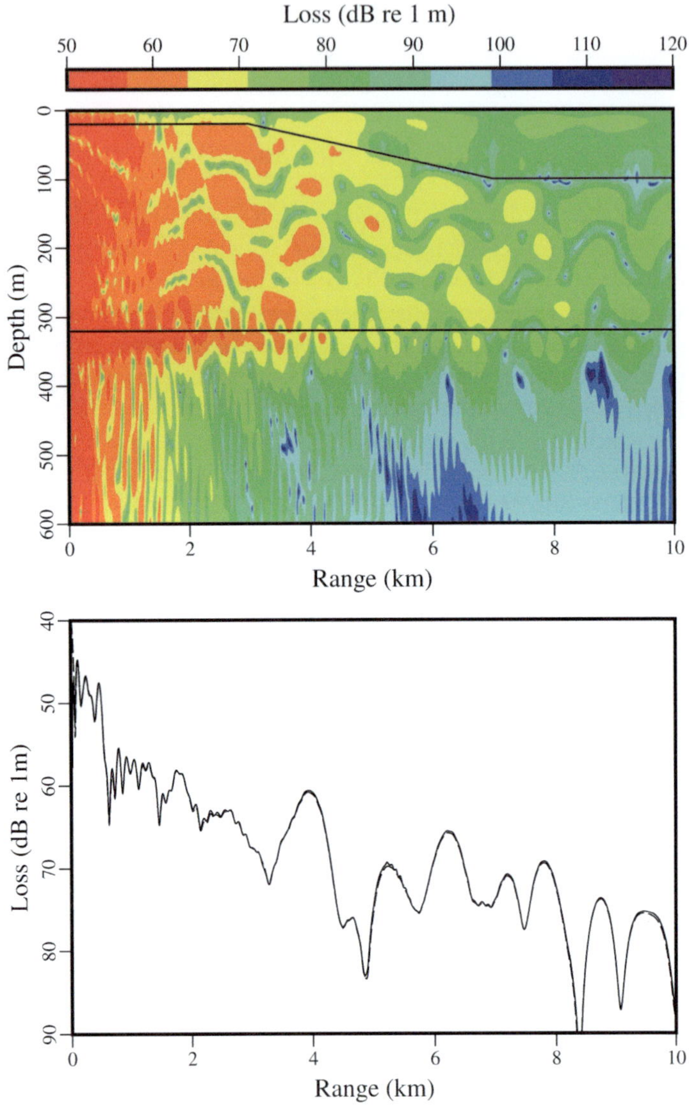

Fig. 3.12 Normal stress transmission loss for a problem involving a source at $z = 300$ m in the water column, which lies below a downward sloping interface with ice and above an elastic basement. The curves in the lower half of the figure correspond to a receiver at $z = 200$ m. The single-scattering solution (solid curve) is in agreement with the finite-element solution (dashed curve)

$$\sigma_{zz} = (\lambda + 4\kappa)\frac{\partial u}{\partial x} + (\lambda + 2\mu - 2\nu)\frac{\partial w}{\partial z}, \tag{3.65}$$

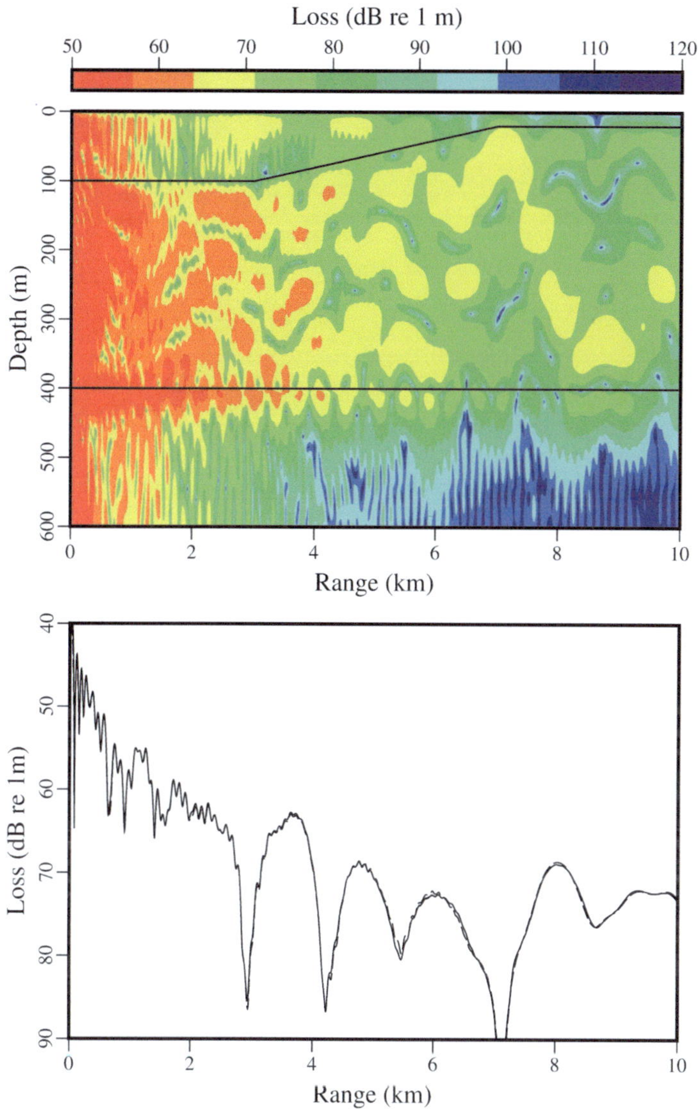

Fig. 3.13 Normal stress transmission loss for a problem involving a source at $z = 380$ m in the water column, which lies below an upward sloping interface with ice and above an elastic basement. The curves in the lower half of the figure correspond to a receiver at $z = 280$ m. The single-scattering solution (solid curve) is in agreement with the finite-element solution (dashed curve)

where λ and μ correspond to the Lamé parameters for the isotropic case $\kappa = \nu = 0$. In this formulation, the coordinate axes are aligned with directions of symmetry. Substituting Eqs. (3.5), (3.64), and (3.65) into Eqs. (3.1) and (3.2), we obtain

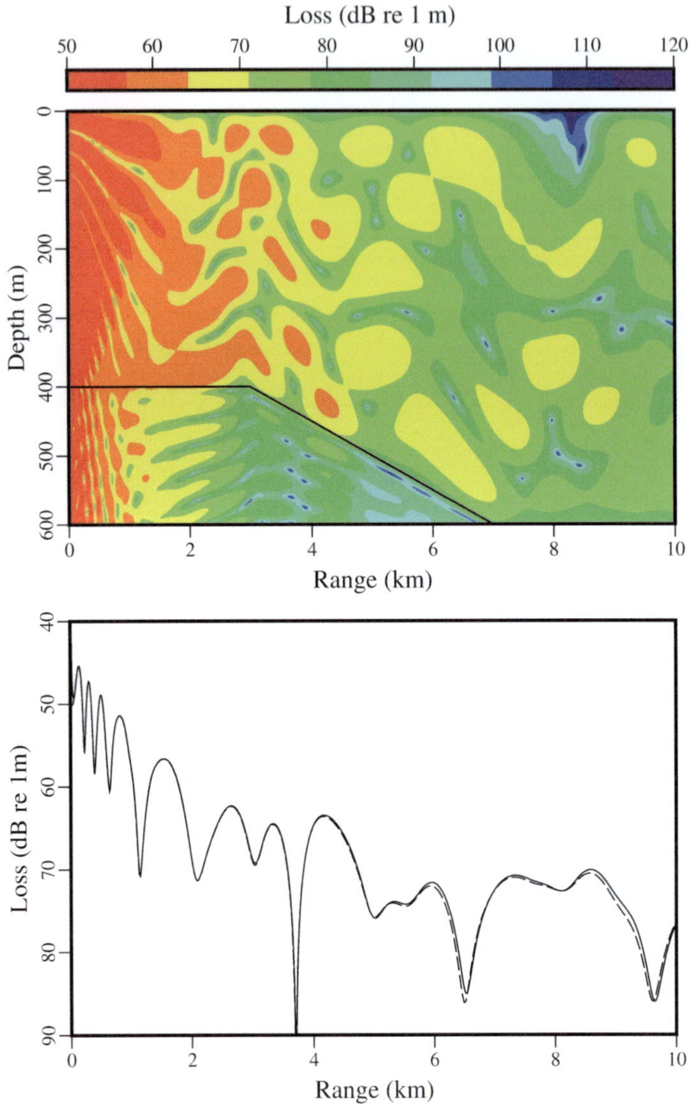

Fig. 3.14 Normal stress transmission loss for a problem involving a source at $z = 380$ m in the water column, which has a downward sloping interface with an elastic basement. The curves in the lower half of the figure correspond to a receiver at $z = 180$ m. The single-scattering solution (solid curve) is in agreement with the finite-element solution (dashed curve)

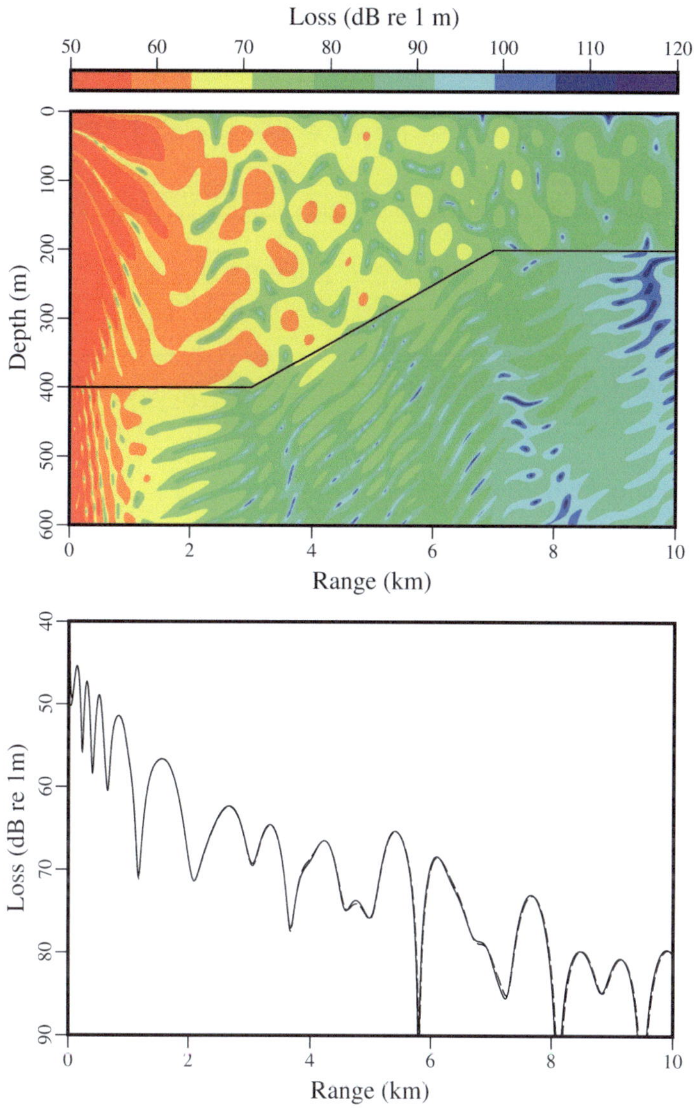

Fig. 3.15 Normal stress transmission loss for a problem involving a source at $z = 380$ m in the water column, which has an upward sloping interface with an elastic basement. The curves in the lower half of the figure correspond to a receiver at $z = 180$ m. The single-scattering solution (solid curve) is in agreement with the finite-element solution (dashed curve)

$$(\lambda+2\mu+2\nu)\frac{\partial^2 u}{\partial x^2}+(\lambda+4\kappa)\frac{\partial^2 w}{\partial x\partial z}+\frac{\partial}{\partial z}\left(\mu\frac{\partial w}{\partial x}\right)+\frac{\partial}{\partial z}\left(\mu\frac{\partial u}{\partial z}\right)+\rho\omega^2 u=0, \quad (3.66)$$

$$\mu\frac{\partial^2 w}{\partial x^2}+\mu\frac{\partial^2 u}{\partial x\partial z}+\frac{\partial}{\partial z}\left((\lambda+4\kappa)\frac{\partial u}{\partial x}\right)+\frac{\partial}{\partial z}\left((\lambda+2\mu-2\nu)\frac{\partial w}{\partial z}\right)+\rho\omega^2 w=0. \quad (3.67)$$

Taking the range derivative of Eq. (3.66) and changing variables, we obtain

$$(\lambda+2\mu+2\nu)\frac{\partial^2 u_x}{\partial x^2}+(\lambda+4\kappa)\frac{\partial^3 w}{\partial x^2\partial z}+\frac{\partial}{\partial z}\left(\mu\frac{\partial^2 w}{\partial x^2}\right)+\frac{\partial}{\partial z}\left(\mu\frac{\partial u_x}{\partial z}\right)+\rho\omega^2 u_x=0, \quad (3.68)$$

$$\mu\frac{\partial^2 w}{\partial x^2}+\mu\frac{\partial u_x}{\partial z}+\frac{\partial}{\partial z}((\lambda+4\kappa)u_x)+\frac{\partial}{\partial z}\left((\lambda+2\mu-2\nu)\frac{\partial w}{\partial z}\right)+\rho\omega^2 w=0. \quad (3.69)$$

These equations are in the form of Eq. (3.17) and may be solved with the same approaches [16].

For the isotropic case, an environment may be specified in terms of λ, μ, and ρ, but it is more natural to use c_p, c_s, and ρ in defining wave propagation problems. The mapping between these alternative sets of parameters is given by Eqs. (3.12) and (3.13). To generalize the mapping to the anisotropic case, we consider a homogeneous medium and plane wave solutions of the form,

$$\begin{pmatrix} u \\ w \end{pmatrix} = \begin{pmatrix} u_0 \\ w_0 \end{pmatrix} \exp\left(i\omega c^{-1}(x\cos\theta + z\sin\theta)\right), \quad (3.70)$$

where u_0, w_0, and c are functions of θ. Substituting into Eqs. (3.66) and (3.67), we obtain

$$det(A) = A_{11}A_{22} - A_{12}A_{21}, \quad (3.71)$$

$$A_{11} = (\lambda + 2\mu + 2\nu)\cos^2\theta + \mu\sin^2\theta - \rho c^2, \quad (3.72)$$

$$A_{12} = A_{21} = (\lambda + \mu + 4\kappa)\cos\theta\sin\theta, \quad (3.73)$$

$$A_{22} = (\lambda + 2\mu - 2\nu)\sin^2\theta + \mu\cos^2\theta - \rho c^2. \quad (3.74)$$

The roots of det(A) correspond to the wave speeds $c_p(\theta)$ and $c_s(\theta)$, where $c_s(\theta) < c_p(\theta)$. For the cases $\theta = 0$ and $\theta = \pi/2$, we obtain

$$\mu = \rho c_s^2(0) = \rho c_s^2(\pi/2), \quad (3.75)$$

$$\lambda + 2\mu = \frac{1}{2}\rho\left(c_p^2(0) + c_p^2(\pi/2)\right), \quad (3.76)$$

$$\nu = \frac{1}{4}\rho\left(c_p^2(0) - c_p^2(\pi/2)\right). \tag{3.77}$$

These equations may be used to determine λ, μ, and ν in terms of ρ and the horizontal and vertical wave speeds. An additional parameter must be provided in order to determine κ. The shear speed has an extremum $\tilde{c}_s$ in the interval $0 < \theta < \pi/2$ (which equals the value at the endpoints if the shear speed is isotropic) that satisfies [12].

$$\lambda + 4\kappa = \rho\sqrt{\left(c_p^2(0) - \tilde{c}_s^2\right)\left(c_p^2(\pi/2) - \tilde{c}_s^2\right)} - \rho\tilde{c}_s^2. \tag{3.78}$$

We consider an example involving a 25 Hz source at $z = 295$ m in an ocean of depth 300 m. Appearing in Fig. 3.16 are the anisotropic wave speeds in the sediment, in which $\rho = 1.5$ g/cm^3, $\sigma_p = 0.2$, and $\sigma_s = 0.4$. We also consider the related isotropic case $c_p = 1675.19$ m/s and $c_s = 800$ m/s, which corresponds to $\kappa = \nu = 0$ and the same values of λ and μ. In the wave number spectra that appear in Fig. 3.17 for these cases, the peaks are in agreement for the first mode but not for the other three modes. Solutions generated with the parabolic equation method and by integrating the wave number spectra appear in Fig. 3.18. There are large differences between the parabolic equation solutions for the anisotropic and related isotropic cases. The parabolic equation and spectral solutions are nearly identical for the anisotropic case. For this problem, acoustic waves that propagate at small angles in the water column couple with compressional waves that propagate at even smaller angles in the sediment and shear waves that propagate at steep angles in the

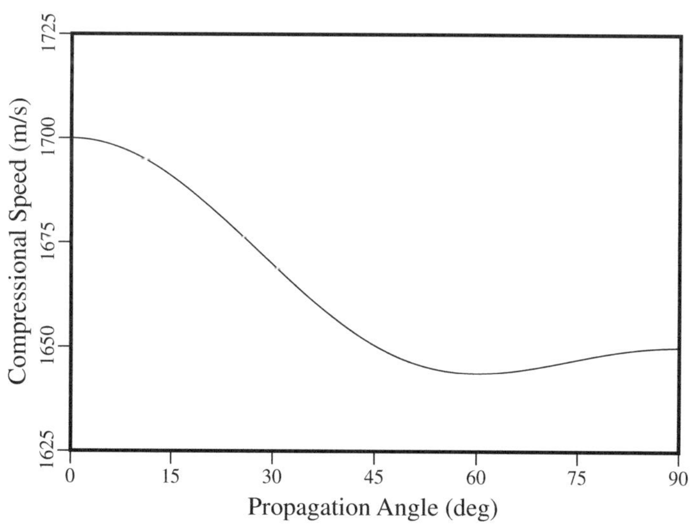

Fig. 3.16 The anisotropic wave speeds for the case $c_p(0) = 1700$ m/s, $c_p(\pi/2) = 1650$ m/s, $c_s(0) = 800$ m/s, and $\tilde{c}_s(0) = 850$ m/s

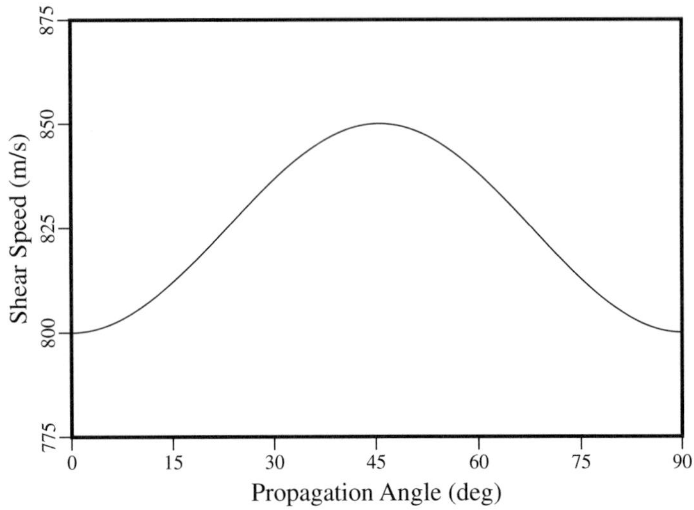

Fig. 3.16 (continued)

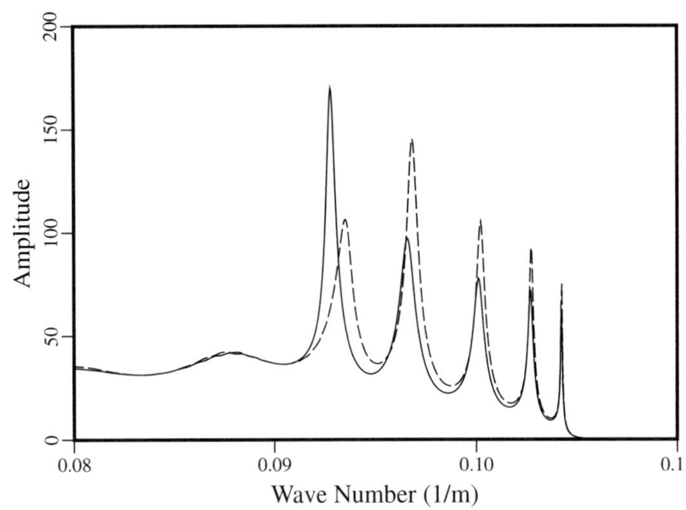

Fig. 3.17 Wave number spectra at $z = 25$ m for examples involving a 25 Hz source in a 300 m deep water column that overlies an anisotropic sediment (solid curve) and a related isotropic sediment (dashed curve)

sediment. This suggests that the case $c_p = 1700$ m/s and $c_s = 850$ m/s would be a better choice for an effective isotropic medium, and in fact the peaks in the wave number spectra are in agreement at $z = 25$ m in Fig. 3.19. This choice is not effective for the Scholte wave, however, and the peaks corresponding that wave are not in agreement at $z = 295$ m in Fig. 3.19.

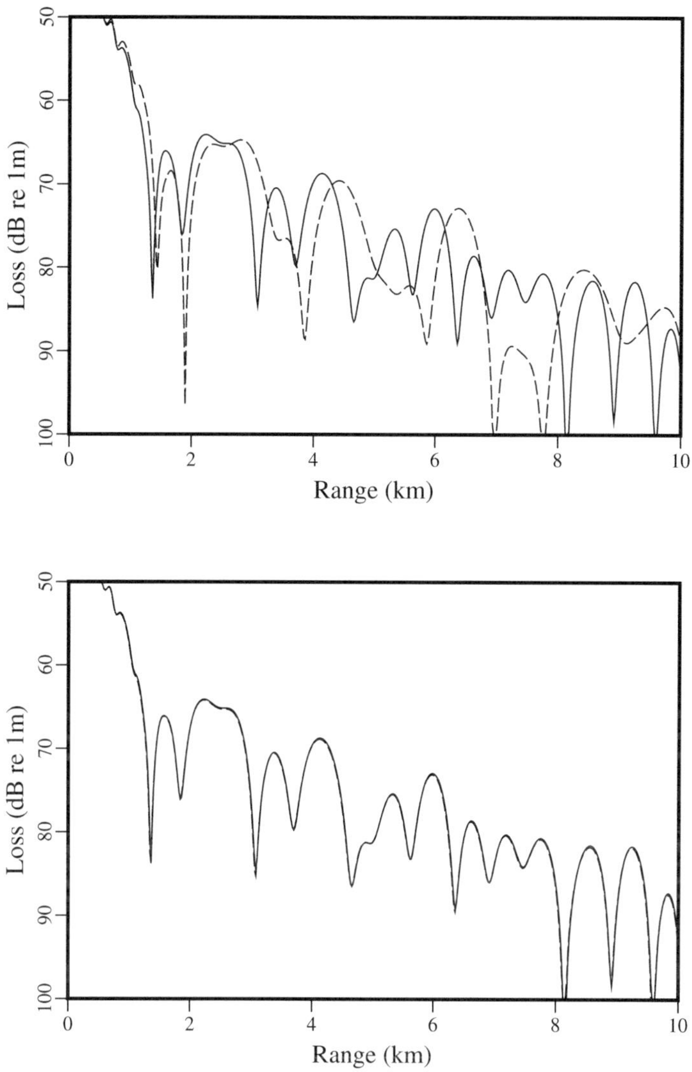

Fig. 3.18 Transmission loss at $z = 25$ m for examples involving a 25 Hz source in a 300 m deep water column. Top: comparison of parabolic equation solutions for anisotropic (solid curve) and isotropic (dashed curve) sediments. Bottom: comparison of parabolic equation (solid curve) and spectral (dashed curve) solutions for the anisotropic case

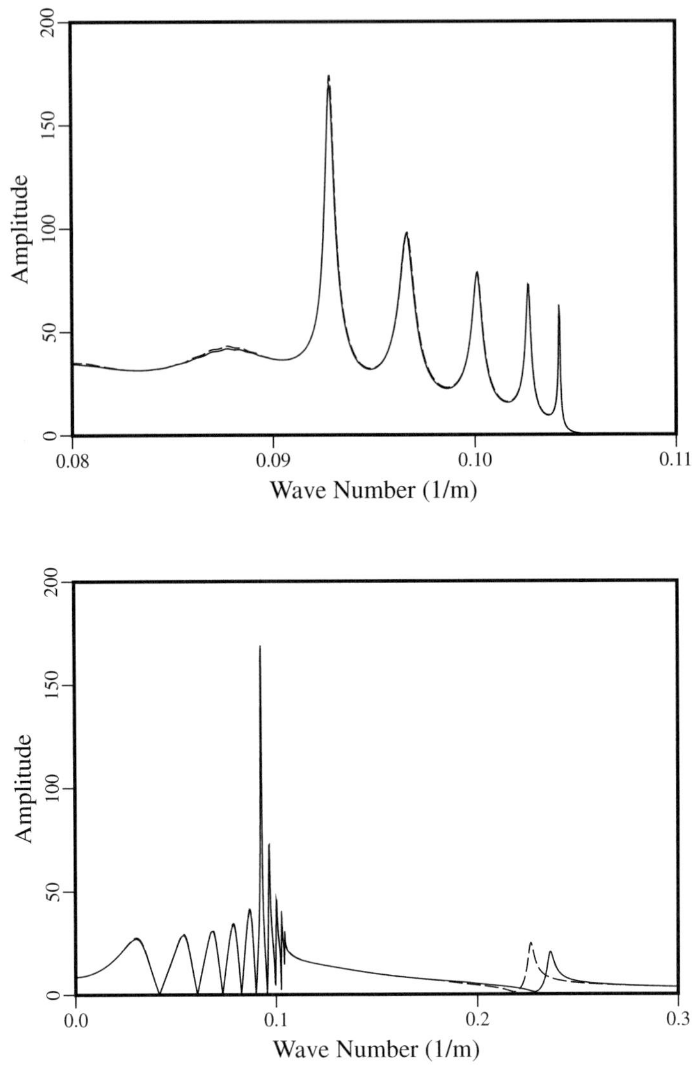

Fig. 3.19 Wave number spectra for examples involving a 25 Hz source in a 300 m deep water column that overlies an anisotropic sediment (solid curves) and a related isotropic sediment (dashed curves). Top: the curves are in agreement at $z = 25$ m. Bottom: the peaks corresponding to the Scholte wave are not in agreement at $z = 295$ m

References

1. H. Kolsky, *Stress Waves in Solids* (Dover, New York, 1963).
2. R.R. Greene, "A high-angle one-way wave equation for seismic wave propagation along rough and sloping interfaces," J. Acoust. Soc. Am. **77**, 1991–1998 (1985).
3. M.D. Collins, "A higher-order parabolic equation for wave propagation in an ocean overlying an elastic bottom," J. Acoust. Soc. Am. **86**, 1459–1464 (1989).

4. M.D. Collins, "Higher-order Padé approximations for accurate and stable elastic parabolic equations with application to interface wave propagation," J. Acoust. Soc. Am. **89**, 1050–1057 (1991).
5. B.T.R. Wetton and G.H. Brooke, "One-way wave equations for seismoacoustic propagation in elastic waveguides," J. Acoust. Soc. Am. **87**, 624–632 (1990).
6. W. Jerzak, W.L. Siegmann, and M.D. Collins, "Modeling Rayleigh and Stoneley waves and other interface and boundary effects with the parabolic equation," J. Acoust. Soc. Am. **117**, 3497–3503 (2005).
7. M.D. Collins and W.L. Siegmann, "Treatment of a sloping fluid-solid interface and sediment layering with the seismo-acoustic parabolic equation," J. Acoust. Soc. Am. **137**, 492–497 (2015).
8. COMSOL Multiphysics® v. 5.2 (COMSOL AB, Stockholm, Sweden).
9. K. Woolfe, M.D. Collins, D.C. Calvo, and W.L. Siegmann, "Seismo-acoustic benchmark problems involving sloping solid-solid interfaces and variable topography," J. Comp. Acoust. **24**, 1650019 (2016).
10. D.A. Outing, W.L. Siegmann, M.D. Collins, and E.K. Westwood, "Generalization of the rotated parabolic equation to variable slopes," J. Acoust. Soc. Am. **120**, 3534–3538 (2006).
11. M.D. Collins, "A single-scattering correction for the seismo-acoustic parabolic equation," J. Acoust. Soc. Am. **131**, 2638–2642 (2012).
12. K. Woolfe, M.D. Collins, D.C. Calvo, and W.L. Siegmann, "Seismo-acoustic benchmark problems involving sloping fluid-solid interfaces," J. Comp. Acoust. **24**, 1650022 (2016).
13. R.T. Bachman, "Acoustic anisotropy in marine sediments and sedimentary rocks," J. Geophys. Res. **84**, 7661–7663 (1979).
14. R.T. Bachman, "Elastic anisotropy in marine sedimentary rocks," J. Geophys. Res. **88**, 539–545 (1983).
15. D.W. Oakley and P.J. Vidmar, "Acoustic reflection from transversely isotropic consolidated sediments," J. Acoust. Soc. Am. **73**, 513–519 (1983).
16. A.J. Fredricks, W.L. Siegmann, and M.D. Collins, "A parabolic equation for anisotropic elastic media," Wave Motion **31**, 139–146 (2000).

Chapter 4
Additional Applications

Contents

4.1 Introduction

The parabolic equation method is widely used in ocean acoustics and seismology. There are many additional applications, including a few that are covered in this chapter. The acoustic wave equation is derived from the equations of fluid mechanics by considering a small amplitude perturbation of a steady state [1]. The same approach may be used to derive linear wave equations that account for the effects of ambient flow and buoyancy. Wind often has significant effects in atmospheric acoustics. A generalization of the adiabatic mode solution is applicable to the case of small Mach number in which the effects of advection can be treated as a perturbation [2]. Internal gravity waves and acousto-gravity waves exist when the effects of buoyancy are significant [3]. A poro-elastic medium is a porous solid with fluid-filled pore spaces. The poro-elastic wave equation was originally derived by Biot using a macroscopic approach [4–7], but it has also been derived as a limiting case of the elastic wave equation for a fluid-solid conglomerate [8]. Poro-elastic media support fast compressional, slow compressional, shear, boundary, and interface waves. The fast wave corresponds to motions of the solid frame and the fluid that are in phase with each other. The slow wave, which corresponds to motions that are out of phase, is strongly attenuated by viscous effects but has been observed

M. D. Collins, W. L. Siegmann, *Parabolic Wave Equations with Applications*,
https://doi.org/10.1007/978-1-4939-9934-7_4

experimentally [9]. One of the applications of the poro-elastic parabolic equation is ocean acoustics problems involving interactions with porous sediments [10].

4.2 Horizontally Advected Acoustic Waves

In deriving a wave equation for advected acoustic waves, we assume that the horizontal components of the wind dominate the vertical component, which is neglected, and that all of the ambient parameters vary gradually in the vertical over the scale of a wavelength and vary much more gradually in the horizontal directions. We define the acoustic pressure p, sound speed c, ambient density ρ, wind velocity $\mathbf{u} = (u_x, u_y, u_z)$, and acoustic velocity $\mathbf{v} = (v_x, v_y, v_z)$. Applying the assumptions to Eq. (15) of Ref. [11], we obtain

$$\rho\nabla\cdot\left(\frac{1}{\rho}\nabla p\right) - \frac{1}{c^2}D_t^2 p + 2\rho\frac{\partial u_x}{\partial z}\frac{\partial v_z}{\partial x} + 2\rho\frac{\partial u_y}{\partial z}\frac{\partial v_z}{\partial y} = 0, \tag{4.1}$$

$$D_t = \frac{\partial}{\partial t} + \mathbf{u}\cdot\nabla. \tag{4.2}$$

We define $\mathbf{U} = c^{-1}\mathbf{u}$, assume that the Mach number $|\mathbf{U}|$ is small, work in the frequency domain, and obtain the approximations,

$$\mathbf{v} \cong -\frac{i}{\omega\rho}\nabla p, \tag{4.3}$$

$$D_t^2 \cong -\omega^2 - 2i\omega\mathbf{u}\cdot\nabla, \tag{4.4}$$

which are valid in the limit of small Mach number. Substituting into Eq. (4.1), we obtain

$$\frac{\partial^2 p}{\partial x^2} + \frac{\partial^2 p}{\partial y^2} + (L+M)p = 0, \tag{4.5}$$

$$Lp = \rho\frac{\partial}{\partial z}\left(\frac{1}{\rho}\frac{\partial p}{\partial z}\right) + k^2 p, \tag{4.6}$$

$$Mp = 2ikU_x\frac{\partial p}{\partial x} + 2ikU_y\frac{\partial p}{\partial y} - 2i\frac{\partial}{\partial z}\left(\frac{U_x}{k}\right)\frac{\partial^2 p}{\partial x\partial z} - 2i\frac{\partial}{\partial z}\left(\frac{U_y}{k}\right)\frac{\partial^2 p}{\partial y\partial z}. \tag{4.7}$$

In the absence of wind, Eq. (4.5) reduces to Eq. (1.7).

We derive a correction to the adiabatic mode solution in Eqs. (2.72) and (2.73) by treating Mp as a perturbation and seeking a solution of the form,

$$p(x,y,z) \cong p_m(x,y)[\zeta_m(z;x,y) + \xi_m(z;x,y)], \tag{4.8}$$

$$L\zeta_m = k_m^2\zeta_m, \tag{4.9}$$

$$\frac{\partial^2 p_m}{\partial x^2} + \frac{\partial^2 p_m}{\partial y^2} + k_m^2 p_m \cong -Q_m p_m, \tag{4.10}$$

where the corrections ξ_m and $Q_m p_m$ are first order in Mach number. Substituting into Eq. (4.5), using Eq. (4.9), and neglecting terms that are second order in Mach number, we obtain

$$p_m(L - k_m^2)\xi_m = \zeta_m Q_m p_m - M(\zeta_m p_m). \tag{4.11}$$

By the Fredholm alternative [12], Eq. (4.11) is solvable if the following condition is satisfied:

$$Q_m p_m = \int \rho^{-1}\zeta_m M(\zeta_m p_m)dz. \tag{4.12}$$

Substituting into Eq. (4.10) and neglecting terms that are second order in Mach number, we obtain

$$\frac{\partial^2 p_m}{\partial x^2} + 2ik_m U_{m,x}\frac{\partial p_m}{\partial x} + \frac{\partial^2 p_m}{\partial y^2} + 2ik_m U_{m,y}\frac{\partial p_m}{\partial y} + k_m^2 p_m = 0, \tag{4.13}$$

$$\mathbf{U}_m = \left(U_{m,x}, U_{m,y}\right) = k_m^{-1}\int \rho^{-1}\zeta_m\left[k\mathbf{U}\zeta_m - \frac{\partial}{\partial z}\left(\frac{\mathbf{U}}{k}\right)\frac{\partial \zeta_m}{\partial z}\right]dz. \tag{4.14}$$

For the case in which $\mathbf{u}$ is independent of depth, $\mathbf{U}_m = C_m^{-1}\left(u_x, u_y\right)$, where the modal phase speed $c_m = \omega/k_m$ and group speed C_m satisfy [2]

$$\frac{1}{c_m C_m} = \int \frac{\zeta_m^2}{\rho c^2}dz. \tag{4.15}$$

Rearranging Eq. (4.13) and neglecting a term that is second order in Mach number, we obtain

$$\left(\frac{\partial}{\partial x} + ik_m U_{m,x}\right)^2 p_m + \frac{\partial^2 p_m}{\partial y^2} + 2ik_m U_{m,y}\frac{\partial p_m}{\partial y} + k_m^2 p_m = 0. \tag{4.16}$$

Factoring the operator in Eq. (4.16) and assuming that outgoing energy dominates, we obtain the parabolic equation,

$$\frac{\partial p_m}{\partial x} = -k_m U_{m,x} p_m + i\left(\frac{\partial^2}{\partial y^2} + 2ik_m U_{m,y}\frac{\partial}{\partial y} + k_m^2\right)^{1/2} p_m. \tag{4.17}$$

For global-scale problems, we obtain the following generalization of Eq. (2.74):

$$\frac{\partial p_m}{\partial r} = -k_m U_{m,r} p_m + i\left(\frac{1}{R_0^2 \sin^2(r/R_0)}\frac{\partial^2}{\partial \theta^2} + \frac{2ik_m U_{m,\theta}}{R_0 \sin(r/R_0)}\frac{\partial}{\partial \theta} + k_m^2\right)^{1/2} p_m, \tag{4.18}$$

where R_0 is the radius of the planet. In order to test Eq. (4.18), we consider a 25 Hz problem for which $R_0 = 12$ km, the source is located at the equator, $c_m = C_m = 1500$ m/s, and the wind blows parallel to lines of constant latitude, with $u_x = 50 \sin(\lambda\pi/9)$ m/s and λ being the latitude in degrees. Appearing in Fig. 4.1 is a comparison of the solution of Eq. (4.18) with rays that were obtained using an approach described in Ref. [2]. As illustrated in Fig. 4.2, the zonal wind pattern causes energy to be horizontally ducted into caustics that propagate away from the source to the east and west. These features and the pattern of caustics on the backside of the planet are similar in the parabolic equation and ray solutions.

The adiabatic mode solution can be used to model the propagation of acoustic waves that are trapped in the sound channel in the atmosphere of Jupiter [13], which is illustrated in Fig. 4.3. The minimum sound speed occurs just above the cloud layers, which favors the observation of the effects of trapped acoustic waves that oscillate up and down in the channel and repeatedly pass through (and interact with) the clouds while propagating outward. We use the zonal winds appearing in Fig. 4.4, which are based on Voyager data [14] that have been interpolated and extrapolated to the poles [2]. Since the available wind speed data are based on motions of the cloud tops, which are near the sound channel axis, we approximate the wind as independent of depth for modes that are confined to a thin layer near the axis. We consider a 0.02 Hz source at the impact latitude of 44° S of Comet Shoemaker-Levy 9 in the atmosphere of Jupiter, for which $R_0 = 71{,}500$ km. At this frequency, there are several trapped modes [2]. Appearing in Fig. 4.5 is the adiabatic mode solution for the fifth mode, which has phase and group speeds of 1009.0 and 908.5 m/s. As in the solution appearing in Fig. 4.1, the zonal winds give rise to caustics to the east and west of the source.

4.3 Buoyancy Effects on Waves in Fluids

We consider acousto-gravity and internal gravity waves in two dimensions. Linearizing the equation of state and the equations for conservation of momentum, mass, and entropy [1], we obtain

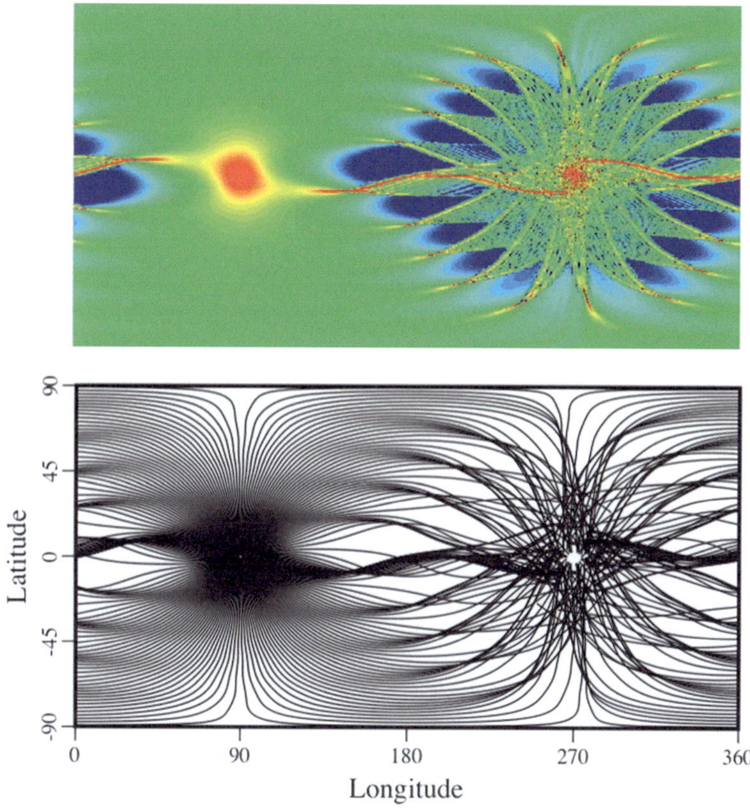

Fig. 4.1 Transmission loss (top) and (rays) bottom for a 25 Hz problem on a planet with a 12 km radius and wind that varies sinusoidally with latitude. The dynamic range is 20 dB. These solutions have similar caustic patterns

$$\left(\frac{\partial^2}{\partial x^2}+k^2\right)p=-i\omega\rho\left(\frac{\partial}{\partial z}+\frac{g}{c^2}\right)w, \tag{4.19}$$

$$\left(N^2-\omega^2\right)w=\frac{i\omega}{\rho}\left(\frac{\partial}{\partial z}-\frac{g}{c^2}\right)p, \tag{4.20}$$

$$N^2=\frac{g}{\rho}\frac{\partial\rho}{\partial z}-\frac{g^2}{c^2}, \tag{4.21}$$

where p and w are perturbations to the ambient state, ρ is the ambient density, g is the acceleration due to gravity, and N is the buoyancy frequency. Using Eq. (4.20) to eliminate w in Eq. (4.19), we obtain the acousto-gravity wave equation [3],

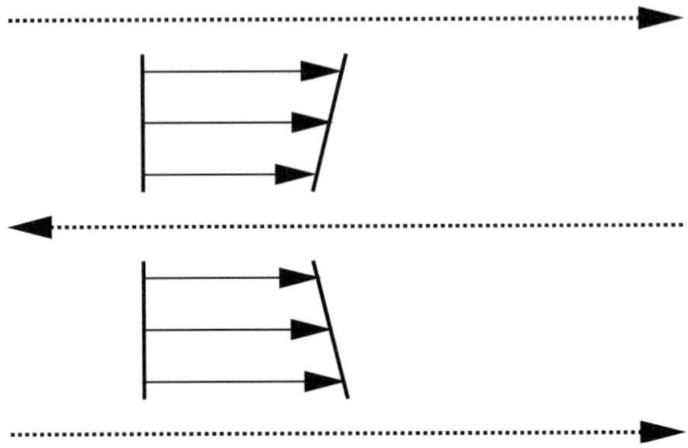

Fig. 4.2 Explanation of the caustics that appear in Fig. 4.1. The dashed lines correspond to the zonal winds, with the arrows indicating the wind direction. Plane wave fronts are initially oriented vertically but bend toward the center by refraction, and the cells between wind reversals act as horizontal waveguides

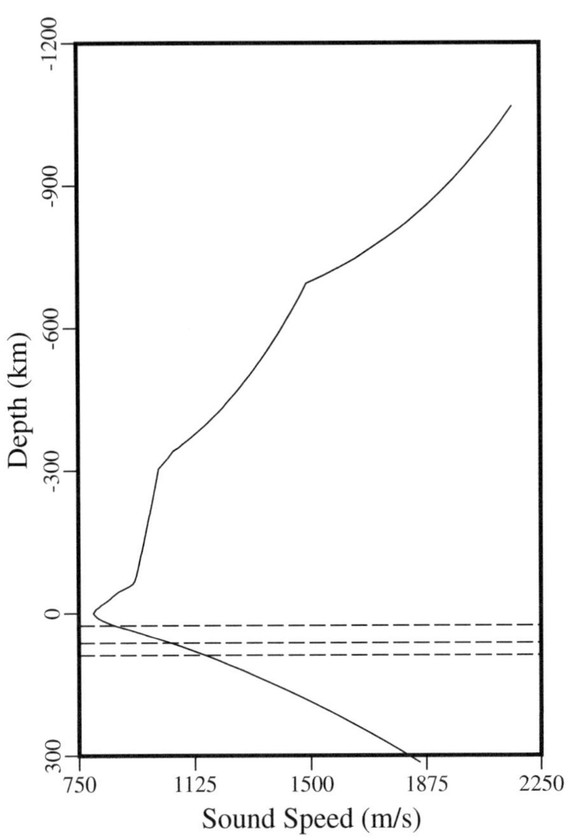

Fig. 4.3 Sound speed of the atmosphere of Jupiter that was obtained from Voyager data [13]. The dashed horizontal lines indicate the locations of (from top to bottom) the ammonia, ammonium hydrosulfide, and water cloud layers. The largest fragments of Comet Shoemaker-Levy 9 are believed to have exploded near the middle cloud layer

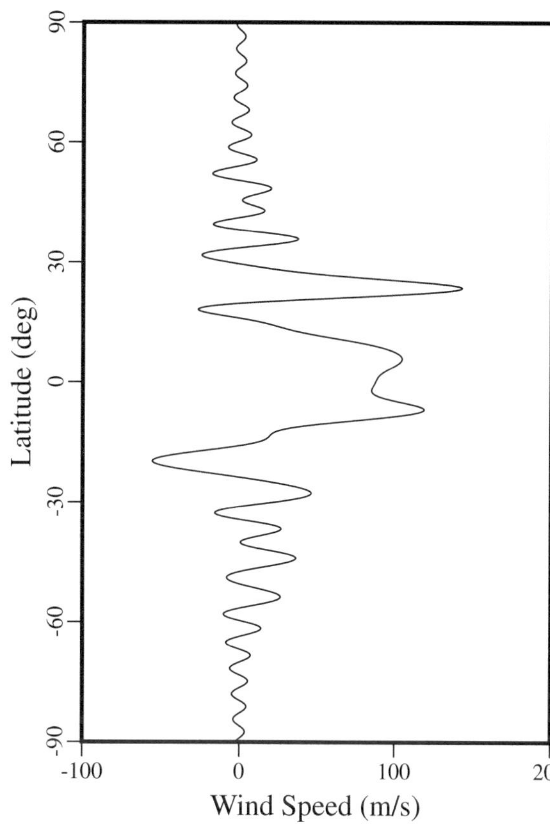

Fig. 4.4 The Jovian zonal winds that were obtained from Voyager data [14] for latitudes within about 60° of the equator and were extrapolated to the poles with decaying sinusoids

$$\frac{\partial^2 p}{\partial x^2}+\rho\frac{\partial}{\partial z}\left(\frac{1}{\rho}\frac{\omega^2}{\omega^2-N^2}\frac{\partial p}{\partial z}\right)+\left(\frac{\omega^2 k^2}{\omega^2-N^2}-\frac{\partial}{\partial z}\left(\frac{gk^2}{\omega^2-N^2}\right)\right)p=0. \tag{4.22}$$

Limiting cases of Eq. (4.22) include the acoustic wave equation,

$$\frac{\partial^2 p}{\partial x^2}+\rho\frac{\partial}{\partial z}\left(\frac{1}{\rho}\frac{\partial p}{\partial z}\right)+k^2 p=0, \tag{4.23}$$

which is obtained by neglecting buoyancy effects, and the gravity wave equation,

$$\frac{\partial^2 p}{\partial x^2}+\rho\frac{\partial}{\partial z}\left(\frac{1}{\rho}\frac{\omega^2}{\omega^2-N^2}\frac{\partial p}{\partial z}\right)p=0, \tag{4.24}$$

which is obtained by assuming that c is much larger than a representative wave speed.

Some of the terms in Eq. (4.22) are singular at depths at which $\omega = N$, but this does not cause any problems in the numerical solution of the parabolic equation,

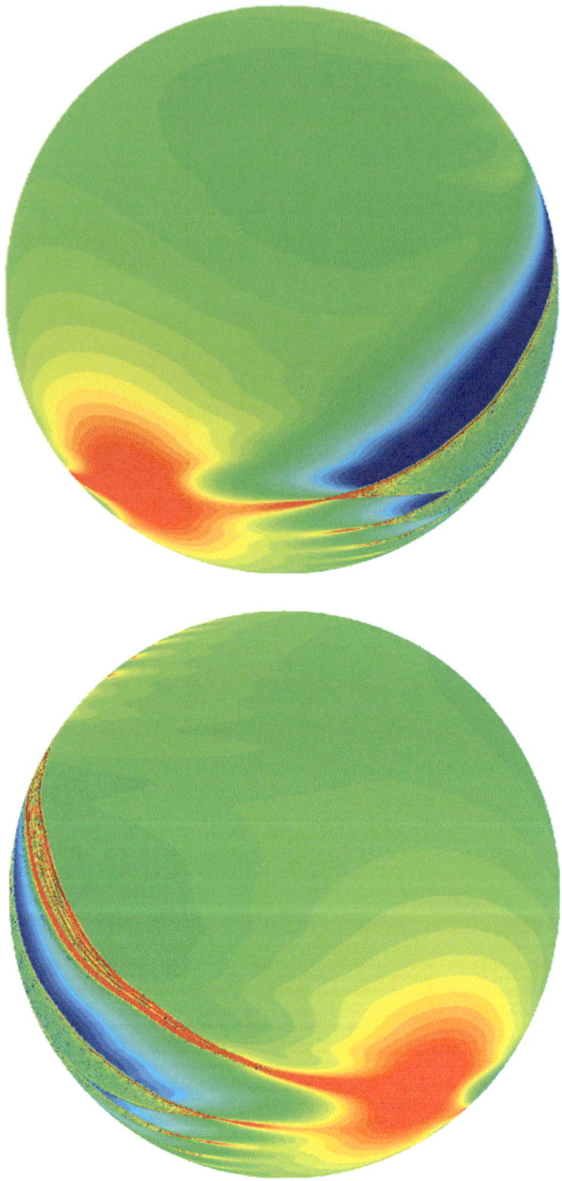

Fig. 4.5 Transmission loss at 0.02 Hz for a source at 44° S that excites the fifth mode. View of Jupiter as it would appear through a telescope that could see acoustic energy. Top: caustics that form to the east of the impact site. Bottom: caustics that appear to the west of the impact site. The dynamic range is 20 dB

$$\frac{\partial p}{\partial x} = ik_0(1+X)^{1/2}p, \tag{4.25}$$

$$k_0^2 X = \rho \frac{\partial}{\partial z}\frac{1}{\rho}\frac{\omega^2}{\omega^2 - N^2}\frac{\partial}{\partial z} + \frac{\omega^2 k^2}{\omega^2 - N^2} - \frac{\partial}{\partial z}\left(\frac{gk^2}{\omega^2 - N^2}\right) - k_0^2. \tag{4.26}$$

Fig. 4.6 A buoyancy frequency ($N/2\pi$) profile that was obtained in the North Pacific Ocean [15]

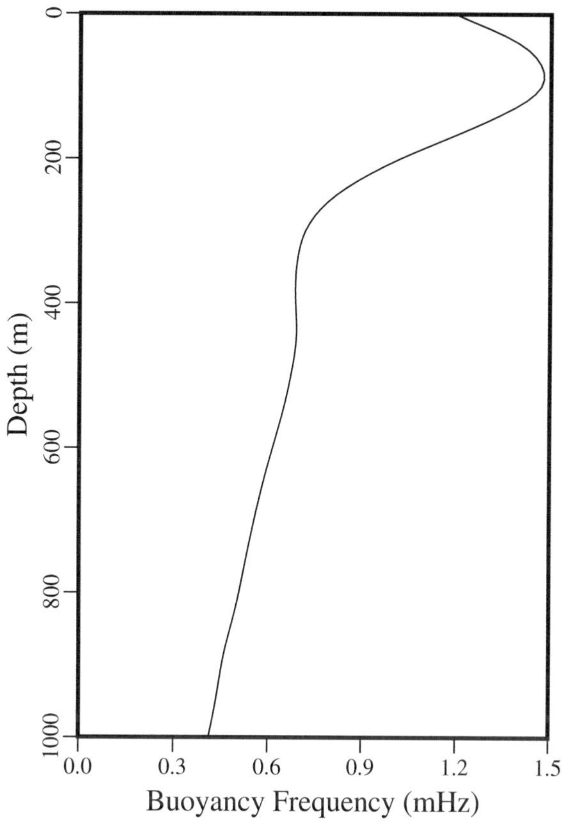

The solution of Eq. (4.25) is based on Eqs. (2.28) and (2.29) and involves equations of the form,

$$\left(1 + \beta_{j,n} X\right) p_j = \left(1 + \gamma_{j,n} X\right) p_{j-1}, \tag{4.27}$$

$$p_0 = \exp\left(i k_0 \Delta x\right) p(x, z), \tag{4.28}$$

$$p_n - p(x + \Delta, z). \tag{4.29}$$

Poles can be avoided by multiplying both sides of Eq. (4.27) by the factor [3],

$$\zeta = \left(\frac{\omega^2}{\omega^2 - N^2}\right)^2. \tag{4.30}$$

Appearing in Fig. 4.6 is a buoyancy profile that was obtained in the North Pacific Ocean [15]. Appearing in Fig. 4.7 is a parabolic equation solution of Eq. (4.24) for a 0.4 mHz problem in an ocean with a sloping seafloor. As described in Ref. [3], an initial condition was obtained by selecting a random combination of modes using the

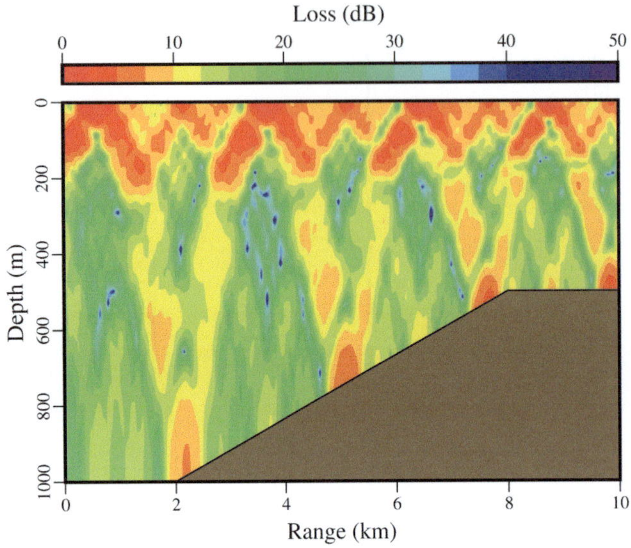

Fig. 4.7 Internal gravity waves at 0.4 mHz that correspond to a realization of the Garrett-Munk spectrum [16] in an ocean with the buoyancy profile in Fig. 4.6

Garrett-Munk spectrum [16]. Energy propagates along curved paths since N depends on depth. There is a relatively large amount of energy in the upper water column because the coefficients of the lower modes (which decay with depth) are relatively large. Appearing in Figs. 4.8, 4.9, and 4.10 are the sound speed, density, and buoyancy frequency in a model atmosphere [17]. As discussed in Ref. [3], the smooth curves were obtained by fitting rational approximations to dashed curves, and artificial attenuation was introduced in the upper atmosphere to prevent reflections from the top boundary of the computation grid. Appearing in Fig. 4.11 are parabolic equation solutions of Eqs. (4.22) and (4.24) for the case of a 3 mHz source at $z = 10$ km. The wave number spectra for both cases appear in Fig. 4.12; the large differences between the gravity and acousto-gravity wave solutions are due to an extra peak in the acousto-gravity wave spectrum, which corresponds to a Lamb wave [1], and other differences for some of the lower modes. Appearing in Fig. 4.13 is a similar problem in which the initial condition is the Lamb wave, which propagates across a mountain range. The variable topography causes mode coupling.

4.4 Poro-Elastic Media

In this section, we consider poro-elastic wave propagation in two dimensions. We use the notation of Ref. [10], where the porosity α is the fraction by volume of the pore spaces, ρ_s is the density of the solid, ρ_f is the density of the fluid, and c_1, c_2, and c_3 are the fast compressional, slow compressional, and shear wave speeds.

Fig. 4.8 A smooth sound speed profile (solid curve) that was obtained by fitting to a standard profile [17] (dashed curve) using the approach described in Ref. [3]

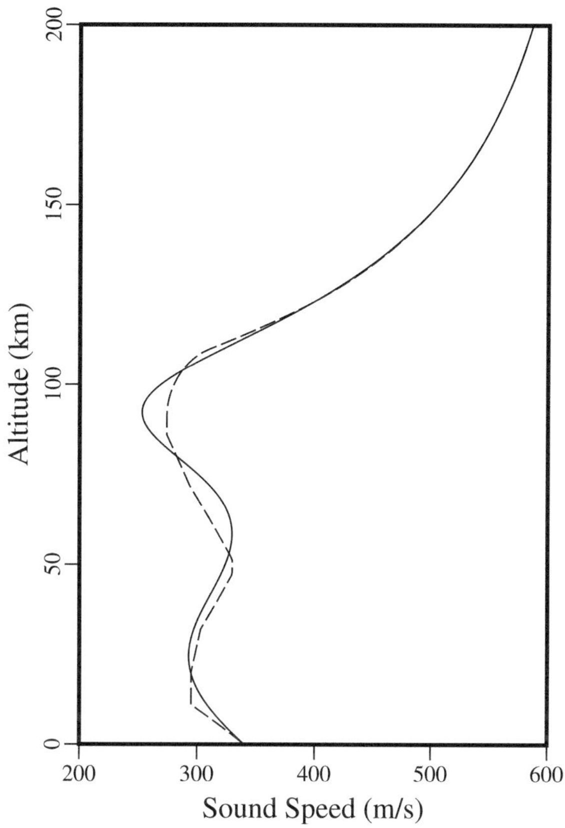

The effects of attenuation are taken into account by allowing the wave speeds to be complex. The dependent variables include the displacement of the porous solid $\mathbf{u}$, the displacement of the fluid $\mathbf{u}_f$, and the relative displacement $\tilde{\mathbf{u}} = \alpha(\mathbf{u}_f - \mathbf{u})$.

4.4.1 The Poro-Elastic Wave Equation

The following generalizations of Eqs. (3.1) and (3.2) correspond to Newton's second law for the combination of the solid and fluid components of a small volume element in two dimensions [4, 7]:

$$\frac{\partial \sigma_{xx}}{\partial x} + \frac{\partial \sigma_{xz}}{\partial z} + (1-\alpha)\rho_s \omega^2 u + \alpha \rho_f \omega^2 u_f = 0, \tag{4.31}$$

$$\frac{\partial \sigma_{xz}}{\partial x} + \frac{\partial \sigma_{zz}}{\partial z} + (1-\alpha)\rho_s \omega^2 w + \alpha \rho_f \omega^2 w_f = 0, \tag{4.32}$$

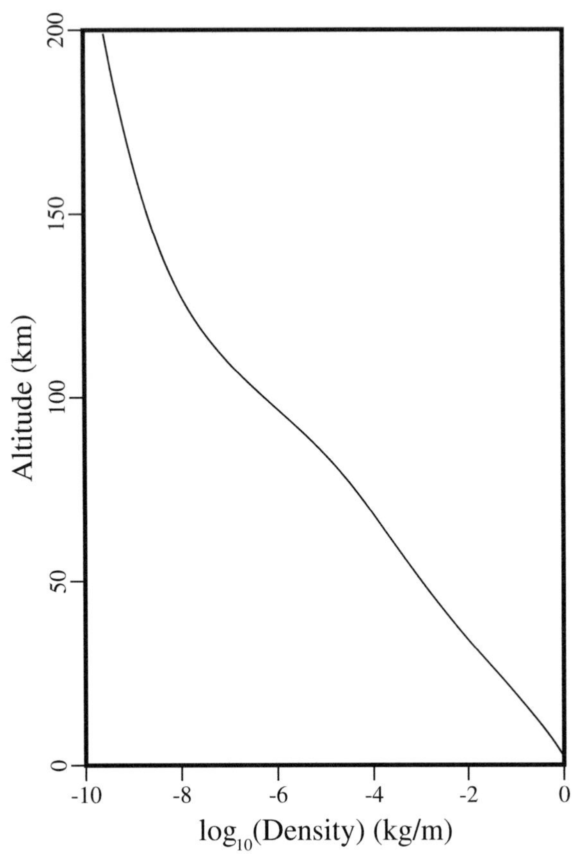

Fig. 4.9 A smooth density profile that was obtained by fitting to a standard profile [17] using the approach described in Ref. [3]

where σ_{xx}, σ_{zz}, and σ_{xz} are the total stresses of the combination of the solid and fluid. The following equations correspond to Newton's second law for the fluid component [4, 7]:

$$\frac{\partial \sigma}{\partial x} + \rho_f \omega^2 u_f + \left(\frac{\rho_f \tau \omega^2}{\alpha} + \frac{i\eta\omega}{\kappa} \right) \tilde{u} = 0, \tag{4.33}$$

$$\frac{\partial \sigma}{\partial z} + \rho_f \omega^2 w_f + \left(\frac{\rho_f \tau \omega^2}{\alpha} + \frac{i\eta\omega}{\kappa} \right) \tilde{w} = 0, \tag{4.34}$$

where $-\sigma$ is the fluid pressure, the terms involving the added mass τ account for the geometry of the pore spaces, and the terms involving the viscosity η and permeability κ correspond to Darcy's law for the flow of a fluid through a porous medium [4]. We apply the following linear relationships between the stresses and distortions [4, 7]:

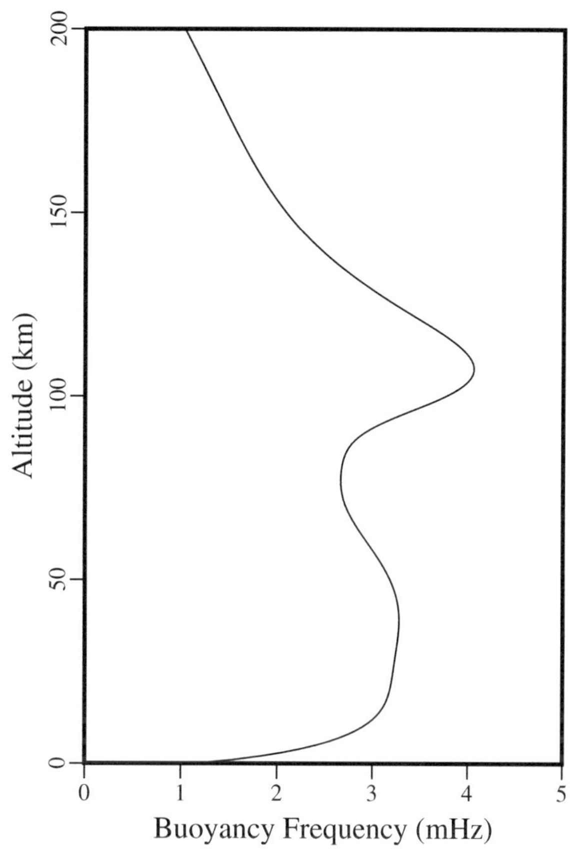

Fig. 4.10 A smooth buoyancy ($N/2\pi$) profile that was obtained by fitting to a standard profile [17] using the approach described in Ref. [3]

$$\sigma_{xx} = \lambda\Delta + 2\mu\,\frac{\partial u}{\partial x} + C\zeta, \tag{4.35}$$

$$\sigma_{zz} = \lambda\Delta + 2\mu\,\frac{\partial w}{\partial z} + C\zeta, \tag{4.36}$$

$$\sigma_{xz} = \mu\left(\frac{\partial u}{\partial z} + \frac{\partial w}{\partial x}\right), \tag{4.37}$$

$$\sigma = C\Delta + M\zeta, \tag{4.38}$$

where $\Delta = \nabla \cdot \mathbf{u}$, $\zeta = \nabla \cdot \tilde{\mathbf{u}}$, and λ, μ, C, and M are the Biot moduli.

Substituting into the equations for Newton's second law, we obtain

$$\frac{\partial}{\partial x}\left(\lambda\Delta + 2\mu\frac{\partial u}{\partial x} + C\zeta\right) + \frac{\partial}{\partial z}\left(\mu\left(\frac{\partial u}{\partial z} + \frac{\partial w}{\partial x}\right)\right) + (1-\alpha)\rho_s\omega^2 u + \alpha\rho_f\omega^2 u_f = 0, \tag{4.39}$$

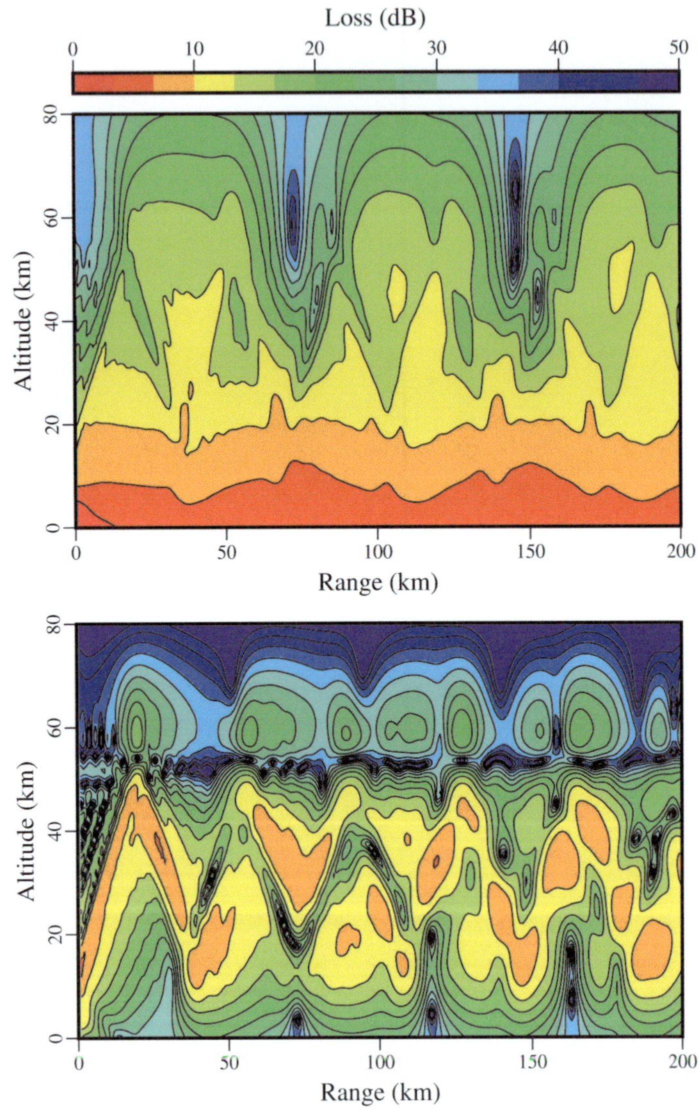

Fig. 4.11 Parabolic equation solutions of the acousto-gravity (top) and gravity (bottom) wave equations for a 3 mHz source at $z = 10$ km. The acousto-gravity wave field is strong near the ground due to the contribution on the Lamb wave. The gravity wave field is weak outside the lower bulge in the buoyancy frequency profile in Fig. 4.10

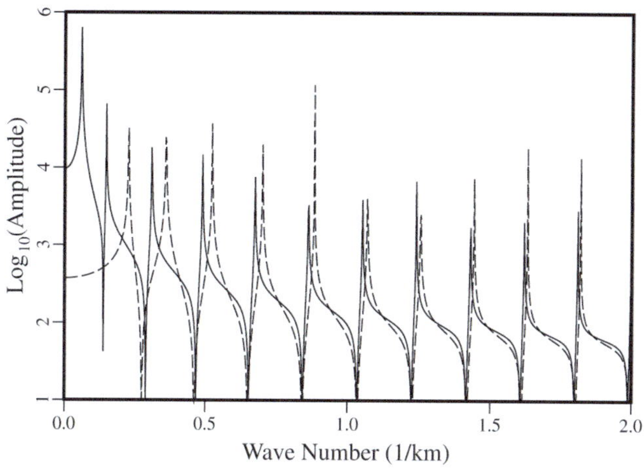

Fig. 4.12 Wave number spectra at $z = 10$ km for the acousto-gravity (solid curve) and gravity (dashed) wave cases that appear in Fig. 4.11. The peaks are nearly aligned for relatively large horizontal wave numbers. There are large differences in the spectra for small horizontal wave numbers, including an extra peak corresponding to the Lamb wave in the acousto-gravity wave spectrum

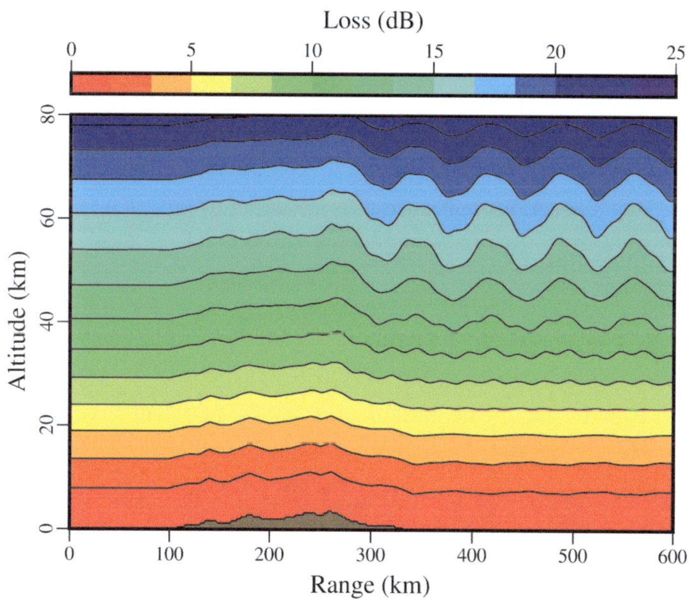

Fig. 4.13 Parabolic equation solution of the acousto-gravity wave equation for a 3 mHz Lamb wave that propagates across a mountain range. To the left of the variable topography, the field consists of a single mode that decays exponentially with altitude. The variable topography causes mode coupling

$$\frac{\partial}{\partial x}\left(\mu\left(\frac{\partial u}{\partial z}+\frac{\partial w}{\partial x}\right)\right)+\frac{\partial}{\partial z}\left(\lambda\Delta+2\mu\frac{\partial w}{\partial z}+C\zeta\right)+(1-\alpha)\rho_s\omega^2 w+\alpha\rho_f\omega^2 w_f$$
$$=0, \tag{4.40}$$

$$\frac{\partial}{\partial x}(C\Delta+M\zeta)+\rho_f\omega^2 u_f+\left(\frac{\rho_f\tau\omega^2}{\alpha}+\frac{i\eta\omega}{\kappa}\right)\tilde{u}=0, \tag{4.41}$$

$$\frac{\partial}{\partial z}(C\Delta+M\zeta)+\rho_f\omega^2 w_f+\left(\frac{\rho_f\tau\omega^2}{\alpha}+\frac{i\eta\omega}{\kappa}\right)\tilde{w}=0. \tag{4.42}$$

For the case of a homogeneous medium, we apply the definition of $\tilde{\mathbf{u}}$ to eliminate $\mathbf{u}_f$ and obtain

$$\mu\nabla^2\mathbf{u}+(\lambda+\mu)\nabla\Delta+C\nabla\zeta+\rho\omega^2\mathbf{u}+\rho_f\omega^2\tilde{\mathbf{u}}=0, \tag{4.43}$$

$$C\nabla\Delta+M\nabla\zeta+\rho_f\omega^2\mathbf{u}+\rho_c\omega^2\tilde{\mathbf{u}}=0, \tag{4.44}$$

where

$$\rho=(1-\alpha)\rho_s+\alpha\rho_f, \tag{4.45}$$

$$\rho_c=\frac{(1+\tau)\rho_f}{\alpha}+\frac{i\eta}{\omega\kappa}. \tag{4.46}$$

In Ref. [4], Biot's equations appear in a form similar to Eqs. (4.43) and (4.44).

4.4.2 The Poro-Elastic Parabolic Equation

Eliminating $\mathbf{u}_f$ and using the definitions of Δ and ρ, we obtain

$$\mu\frac{\partial^2 u}{\partial x^2}+\mu\frac{\partial^2 u}{\partial z^2}+\rho\omega^2 u+(\lambda+\mu)\frac{\partial\Delta}{\partial x}+\frac{\partial\mu}{\partial z}\frac{\partial u}{\partial z}+\frac{\partial\mu}{\partial z}\frac{\partial w}{\partial x}+C\frac{\partial\zeta}{\partial x}+\rho_f\omega^2\tilde{u}$$
$$=0, \tag{4.47}$$

$$\mu\frac{\partial^2 w}{\partial x^2}+\mu\frac{\partial^2 w}{\partial z^2}+\rho\omega^2 w+(\lambda+\mu)\frac{\partial\Delta}{\partial z}+2\frac{\partial\mu}{\partial z}\frac{\partial w}{\partial z}+\frac{\partial\lambda}{\partial z}\Delta+\frac{\partial}{\partial z}(C\zeta)+\rho_f\omega^2\tilde{w}$$
$$=0, \tag{4.48}$$

$$C\frac{\partial\Delta}{\partial x}+M\frac{\partial\zeta}{\partial x}+\rho_f\omega^2u+\rho_c\omega^2\tilde{u}=0, \tag{4.49}$$

$$\frac{\partial}{\partial z}(C\Delta)+\frac{\partial}{\partial z}(M\zeta)+\rho_f\omega^2w+\rho_c\omega^2\tilde{w}=0. \tag{4.50}$$

For the case $\alpha=0$, Eqs. (4.47) and (4.48) reduce to Eqs. (3.6) and (3.7). Differentiating Eq. (4.47) with respect to x, differentiating Eq. (4.18) with respect to z, and summing, we obtain

$$\begin{aligned}&(\lambda+2\mu)\frac{\partial^2\Delta}{\partial x^2}+(\lambda+2\mu)\frac{\partial^2\Delta}{\partial z^2}+\rho\omega^2\Delta+2\frac{\partial\mu}{\partial z}\frac{\partial^2 w}{\partial x^2}+\omega^2\frac{\partial\rho}{\partial z}w+\left(\frac{\partial\lambda}{\partial z}+2\frac{\partial\mu}{\partial z}\right)\frac{\partial\Delta}{\partial z}\\&+\frac{\partial}{\partial z}\left(\frac{\partial\lambda}{\partial z}\Delta\right)+2\frac{\partial}{\partial z}\left(\frac{\partial\mu}{\partial z}\frac{\partial w}{\partial z}\right)+C\frac{\partial^2\zeta}{\partial x^2}+\frac{\partial^2}{\partial z^2}(C\zeta)+\rho_f\omega^2\zeta+\omega^2\frac{\partial\rho_f}{\partial z}\tilde{w}=0.\end{aligned} \tag{4.51}$$

Differentiating Eq. (4.49) with respect to x, differentiating Eq. (4.50) with respect to z, and summing, we obtain

$$\begin{aligned}&C\frac{\partial^2\Delta}{\partial x^2}+\frac{\partial^2}{\partial z^2}(C\Delta)+\rho_f\omega^2\Delta+M\frac{\partial^2\zeta}{\partial x^2}+\frac{\partial^2}{\partial z^2}(M\zeta)+\rho_c\omega^2\zeta+\omega^2\frac{\partial\rho_f}{\partial z}w\\&+\omega^2\frac{\partial\rho_c}{\partial z}\tilde{w}=0.\end{aligned} \tag{4.52}$$

Using Eq. (4.50) to eliminate $\tilde{w}$ from Eqs. (4.48), (4.51), and (4.52), we obtain a system of equations of the form,

$$\left(L\frac{\partial^2}{\partial x^2}+M\right)\mathbf{U}=0, \tag{4.53}$$

$$\mathbf{U}=\begin{pmatrix}\Delta\\ w\\ \zeta\end{pmatrix}, \tag{4.54}$$

where L and M are 3×3 depth operators. In going from Eqs. (4.43) and (4.44) to Eq. (4.53), a redundant equation in Biot's original formulation is eliminated by reducing from four equations in four variables to three equations in three variables. Factoring Eq. (4.53) and assuming that outgoing energy dominates incoming energy, we obtain the poro-elastic parabolic equation,

$$\frac{\partial\mathbf{U}}{\partial x}=i\left(L^{-1}M\right)^{1/2}\mathbf{U}. \tag{4.55}$$

This equation is valid for problems involving continuous depth variations in the material properties within a poro-elastic layer. Horizontal interfaces may be handled by applying interface conditions.

There are four conditions at an interface between poro-elastic layer A and fluid layer B. We cast these equations in a form suitable for Eq. (4.53) and the parabolic equation method. Balance of fluid flow corresponds to

$$\left((1-\alpha)w+\alpha w_f\right)_A = w_B. \tag{4.56}$$

Substituting for the fluid displacements and dropping the subscripts, we obtain

$$w+\tilde{w} = \frac{1}{\rho_f\omega^2}\frac{\partial p}{\partial z}. \tag{4.57}$$

This condition may be implemented using Eq. (4.50) to eliminate $\tilde{w}$. Continuity of normal stress corresponds to

$$\lambda\Delta + 2\mu\frac{\partial w}{\partial z} + C\zeta = -p. \tag{4.58}$$

Continuity of fluid pressure corresponds to

$$C\Delta + M\zeta = -p. \tag{4.59}$$

The following condition for vanishing tangential stress is obtained using Eq. (4.32) and the definition of $\tilde{\mathbf{u}}$:

$$\frac{\partial}{\partial z}\left(\lambda\Delta + 2\mu\frac{\partial w}{\partial z} + C\zeta\right) + \rho\omega^2 w + \rho_f\omega^2\tilde{w} = 0. \tag{4.60}$$

There are five conditions at an interface between poro-elastic layer A and elastic layer B. The following conditions correspond to continuity of the solid displacements:

$$w_A = w_B, \tag{4.61}$$

$$\left(\Delta - \frac{\partial w}{\partial z}\right)_A = (u_x)_B. \tag{4.62}$$

The following conditions correspond to continuity of the normal and tangential stresses:

$$\left(\lambda\Delta + 2\mu\frac{\partial w}{\partial z} + C\zeta\right)_A = \left(\lambda u_x + (\lambda + 2\mu)\frac{\partial w}{\partial z}\right)_B, \tag{4.63}$$

$$\left(\frac{\partial}{\partial z}\left(\lambda\Delta + 2\mu\frac{\partial w}{\partial z} + C\zeta\right) + \rho\omega^2 w + \rho_f\omega^2\tilde{w}\right)_A$$
$$= \left(\frac{\partial}{\partial z}\left(\lambda u_x + (\lambda + 2\mu)\frac{\partial w}{\partial z}\right) + \rho\omega^2 w\right)_B. \tag{4.64}$$

Since fluid does not flow into the elastic layer, the remaining boundary condition is

$$\tilde{w}_A = 0. \tag{4.65}$$

In the microscopic limit of elasticity in which the poro-elastic layer is treated as a fluid-solid conglomerate, fluid pressure in the poro-elastic layer balances normal stress in the elastic layer. In Biot's macroscopic theory of poro-elasticity, the treatment of the interface is based on Saint-Venant's principle for forces applied to an interface [18]. The elastic layer contains a boundary layer that is thin relative to a wavelength but thick relative to a pore space. Outside the boundary layer, the stresses are independent of microscopic redistributions of the forces on the interface. Coupling with a poro-elastic layer is handled by assigning the values of the stresses just outside the boundary layer to the boundary. The effective normal stress in the elastic layer balances the total normal stress in the poro-elastic layer but not the fluid pressure.

4.4.3 Wave Speeds and Mappings Between Variables

In a homogeneous layer, compressional and shear wave solutions may be expressed in terms of potentials. Compressional wave solutions are of the form,

$$(u, w) = \left(\frac{\partial\phi_j}{\partial x}, \frac{\partial\phi_j}{\partial z}\right), \tag{4.66}$$

$$(\tilde{u}, \tilde{w}) = \left(A_j\frac{\partial\phi_j}{\partial x}, A_j\frac{\partial\phi_j}{\partial z}\right), \tag{4.67}$$

$$\frac{\partial^2\phi_j}{\partial x^2} + \frac{\partial^2\phi_j}{\partial z^2} + k_j^2\phi_j = 0, \tag{4.68}$$

where $j = 1$ for the fast wave, $j = 2$ for the slow wave, ϕ_j is the potential, $k_j = \omega/c_j$ is the wave number, c_j is the wave speed, and A_j is an amplitude ratio. Substituting into Eqs. (4.47)–(4.50), we obtain

$$\begin{pmatrix} \lambda + 2\mu - \rho c_j^2 & C - \rho_f c_j^2 \\ C - \rho_f c_j^2 & M - \rho_c c_j^2 \end{pmatrix} \begin{pmatrix} 1 \\ A_j \end{pmatrix} = 0, \tag{4.69}$$

$$A_j = -\frac{\lambda + 2\mu - \rho c_j^2}{C - \rho_f c_j^2} = -\frac{C - \rho_f c_j^2}{M - \rho_c c_j^2}. \tag{4.70}$$

Requiring the determinant of the matrix in Eq. (4.69) to vanish, we obtain

$$E c_j^4 - F c_j^2 + G = 0, \tag{4.71}$$

$$E = \rho \rho_c - \rho_f^2, \tag{4.72}$$

$$F = \rho_c(\lambda + 2\mu) + \rho M - 2\rho_f C, \tag{4.73}$$

$$G = (\lambda + 2\mu)M - C^2. \tag{4.74}$$

The wave speeds correspond to the roots,

$$c_1^2 = \frac{F + \sqrt{F^2 - 4EG}}{2E}, \tag{4.75}$$

$$c_2^2 = \frac{F - \sqrt{F^2 - 4EG}}{2E}. \tag{4.76}$$

Shear wave solutions are of the form,

$$(u, w) = \left(-\frac{\partial \phi_3}{\partial z}, \frac{\partial \phi_3}{\partial x} \right), \tag{4.77}$$

$$(\tilde{u}, \tilde{w}) = \left(-A_3 \frac{\partial \phi_3}{\partial z}, A_3 \frac{\partial \phi_3}{\partial x} \right), \tag{4.78}$$

$$\frac{\partial^2 \phi_3}{\partial x^2} + \frac{\partial^2 \phi_3}{\partial z^2} + k_3^2 \phi_3 = 0, \tag{4.79}$$

where ϕ_3 is the potential, $k_3 = \omega/c_3$ is the wave number, c_3 is the wave speed, and A_3 is an amplitude ratio. Substituting into Eqs. (4.47)–(4.50), we obtain

$$\begin{pmatrix} \rho c_3^2 - \mu & \rho_f c_3^2 \\ \rho_f c_3^2 & \rho_c c_3^2 \end{pmatrix} \begin{pmatrix} 1 \\ A_3 \end{pmatrix} = 0, \tag{4.80}$$

$$A_3 = -\frac{\rho_f}{\rho_c} = -\frac{\rho c_3^2 - \mu}{\rho_f c_3^2}, \tag{4.81}$$

$$c_3^2 = \frac{\mu \rho_c}{\rho \rho_c - \rho_f^2}. \tag{4.82}$$

In the low-frequency limit, the expressions in Eqs. (4.75) and (4.82) reduce to the elastic case $\rho c_1^2 = \lambda + 2\mu$ and $\rho c_3^2 = \mu$.

As discussed in Ref. [10], it is possible to invert the mapping from the Biot moduli to the complex wave speeds so that a poro-elastic medium may be specified in terms of the wave speeds and attenuations, which is typically the preferred approach for wave propagation problems. An alternative approach would be to replace the slow wave speed and attenuation, for which estimates are unlikely to be available, with the bulk modulus K_r of the grains that form the porous medium (estimates of this quantity are available for various materials) and the ratio of the viscosity and permeability (which would probably be the most uncertain quantity). It is also convenient to map between different choices for the dependent variables. Since Δ and ζ are independent of the shear wave, the compressional potentials may be obtained from the linear relationship,

$$\begin{pmatrix} \Delta \\ \zeta \end{pmatrix} = -k_1^2 \phi_1 \begin{pmatrix} 1 \\ A_1 \end{pmatrix} - k_2^2 \phi_2 \begin{pmatrix} 1 \\ A_2 \end{pmatrix}. \tag{4.83}$$

From Eqs. (4.48), (4.77)–(4.79), we obtain

$$k_3^2 \mu \frac{\partial \phi_3}{\partial x} = (\lambda + 2\mu) \frac{\partial \Delta}{\partial z} + \rho \omega^2 w + C \frac{\partial \zeta}{\partial z} + \rho_f \omega^2 \tilde{w}. \tag{4.84}$$

From Eqs. (4.50), (4.55), and (4.84) and the definitions of Δ and ζ, we obtain

$$\begin{pmatrix} u \\ k_3^2 \mu \phi_3 \\ \tilde{u} \end{pmatrix} = -iP \left(L^{-1} M\right)^{-1/2} \mathbf{U}, \tag{4.85}$$

$$P = \begin{pmatrix} 1 & -\dfrac{\partial}{\partial z} & 0 \\ \left(\lambda + 2\mu - \dfrac{\rho_f C}{\rho_c}\right) \dfrac{\partial}{\partial z} & \left(\rho - \dfrac{\rho_f^2}{\rho_c}\right) \omega^2 & \left(C - \dfrac{\rho_f M}{\rho_c}\right) \dfrac{\partial}{\partial z} \\ \dfrac{C}{\rho_c \omega^2} \dfrac{\partial^2}{\partial z^2} & \dfrac{\rho_f}{\rho_c} \dfrac{\partial}{\partial z} & 1 + \dfrac{M}{\rho_c \omega^2} \dfrac{\partial^2}{\partial z^2} \end{pmatrix}. \tag{4.86}$$

4.4.4 Source Conditions

When a source is located in a fluid or solid layer, the approaches described in Sect. 3.4 may be used to generate an initial condition. In order to account for a source in a homogeneous poro-elastic layer that excites only one type of compressional wave, we introduce source terms on the right sides of Eqs. (4.51) and (4.52), apply Eqs. (4.66) and (4.67), and obtain

$$(\lambda+2\mu+A_jC)\frac{\partial^2\Delta}{\partial x^2}+(\lambda+2\mu+A_jC)\frac{\partial^2\Delta}{\partial z^2}+(\rho+A_j\rho_f)\omega^2\Delta=\delta(x)\delta(z-z_0), \quad (4.87)$$

$$(C+A_jM)\frac{\partial^2\Delta}{\partial x^2}+(C+A_jM)\frac{\partial^2\Delta}{\partial z^2}+(\rho_f+A_j\rho_c)\omega^2\Delta=R_j\delta(x)\delta(z-z_0), \quad (4.88)$$

where R_j is the relative source amplitude. Comparing the coefficients of Eqs. (4.87) and (4.88), we obtain

$$R_j=\frac{C+A_jM}{\lambda+2\mu+A_jC}=\frac{\rho_f+A_j\rho_c}{\rho+A_j\rho_f}. \quad (4.89)$$

In order to prevent the excitement of a shear wave, we apply the condition,

$$\frac{\partial^2\phi_3}{\partial x^2}+\frac{\partial^2\phi_3}{\partial z^2}+k_3^2\phi_3=0. \quad (4.90)$$

Integrating over a small interval about $x=0$, we obtain

$$\lim_{x\to 0^+}\frac{\partial\phi_3}{\partial x}=0. \quad (4.91)$$

Substituting Eq. (4.84) into Eq. (4.91), using Eq. (4.50) to eliminate $\tilde{w}$, applying Eqs. (4.87) and (4.88), and using Eq. (4.55) to eliminate the range derivative, we obtain

$$\lim_{x\to 0^+}Q\left(L^{-1}M\right)^{1/2}\mathbf{U}=\begin{pmatrix}\delta(z-z_0)\\ 0\\ R_j\delta(z-z_0)\end{pmatrix}, \quad (4.92)$$

$$Q = \begin{pmatrix} 1 & 0 & 0 \\ \left(\lambda + 2\mu - \frac{\rho_f C}{\rho_c}\right)\frac{\partial}{\partial z} & \left(\rho - \frac{\rho_f^2}{\rho_c}\right)\omega^2 & \left(C - \frac{\rho_f M}{\rho_c}\right)\frac{\partial}{\partial z} \\ 0 & 0 & 1 \end{pmatrix}. \tag{4.93}$$

Performing elimination steps on the middle row of Q as in Sect. 3.4.2, we obtain

$$\lim_{x\to 0^+} \left(L^{-1}M\right)^{1/2}\mathbf{U} = \begin{pmatrix} \delta(z-z_0) \\ -S_j\delta'(z-z_0) \\ R_j\delta(z-z_0) \end{pmatrix}, \tag{4.94}$$

$$S_j = \frac{\rho_c(\lambda + 2\mu) - \rho_f C + R_j\left(\rho_c C - \rho_f M\right)}{\left(\rho\rho_c - \rho_f^2\right)\omega^2}. \tag{4.95}$$

Using Eq. (4.55) to march the field out a short distance in range, we obtain the self-starter,

$$\mathbf{U}(x_0, z) = \left(L^{-1}M\right)^{-1/2} \exp\left(ix_0\left(L^{-1}M\right)^{1/2}\right) \begin{pmatrix} \delta(z-z_0) \\ -S_j\delta'(z-z_0) \\ R_j\delta(z-z_0) \end{pmatrix}. \tag{4.96}$$

For the case of a shear source in a homogeneous poro-elastic layer, we obtain

$$\lim_{x\to 0^+} Q\left(L^{-1}M\right)^{1/2}\mathbf{U} = \begin{pmatrix} 0 \\ k_3^2\mu\delta(z-z_0) \\ 0 \end{pmatrix}, \tag{4.97}$$

$$\mathbf{U}(x_0, z) = \exp\left(ix_0\left(L^{-1}M\right)^{1/2}\right) \begin{pmatrix} 0 \\ S_3\delta(z-z_0) \\ 0 \end{pmatrix}, \tag{4.98}$$

$$S_3 = \frac{k_3^2\rho_c\mu}{\rho\rho_c - \rho_f^2}. \tag{4.99}$$

Appearing in Figs. 4.14 and 4.15 are parabolic equation solutions for an example involving an array of 25 Hz shear sources that are phased to excite a beam that propagates upward toward the surface at an angle of -35° relative to the horizontal.

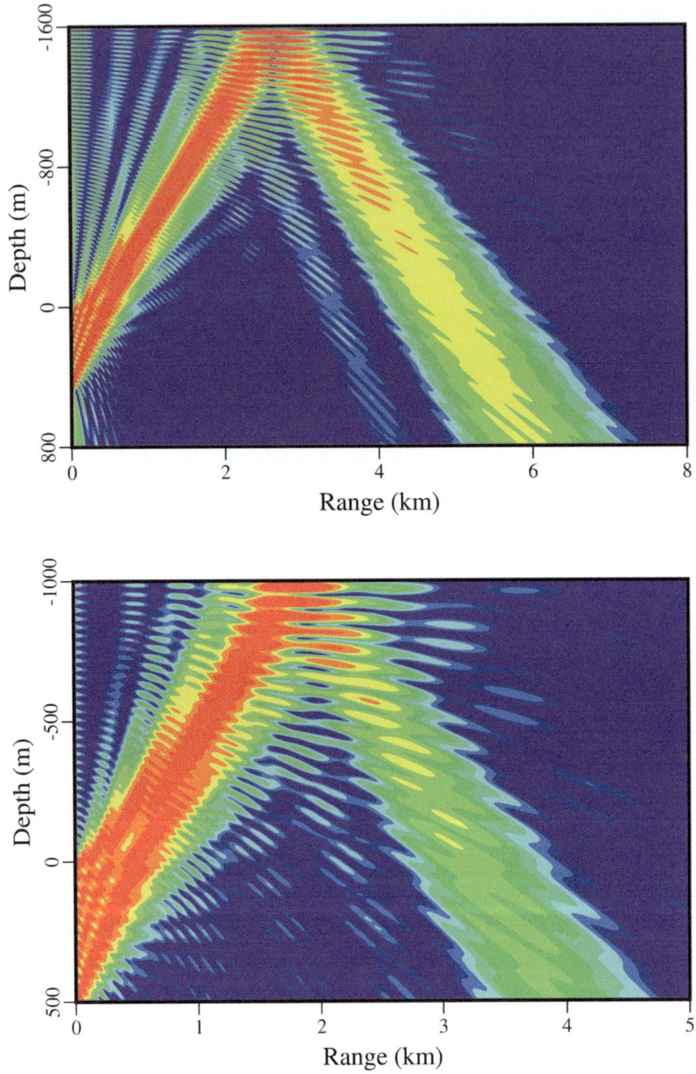

Fig. 4.14 Shear transmission loss (20 dB dynamic range for each plot) for poro-elastic (top) and elastic (bottom) lossless half spaces that contain arrays of twenty shear sources at 25 Hz. For the poro-elastic case, $c_1 = 2400$ m/s, $c_2 = 1000$ m/s, and $c_3 = 1200$ m/s, and the jth source is at a depth of $z_j = (j - 1)24$ m (half wavelength spacing) and has the phase factor $\exp(-ik_3 z_j \sin 35°)$. The elastic case is as in Fig. 3.7

The self-starter properly excites the shear wave beam but does not excite either of the compressional waves. The main beam is incident beyond the critical angle for coupling to the fast compressional wave, but the more steeply propagating side beams excite fast waves.

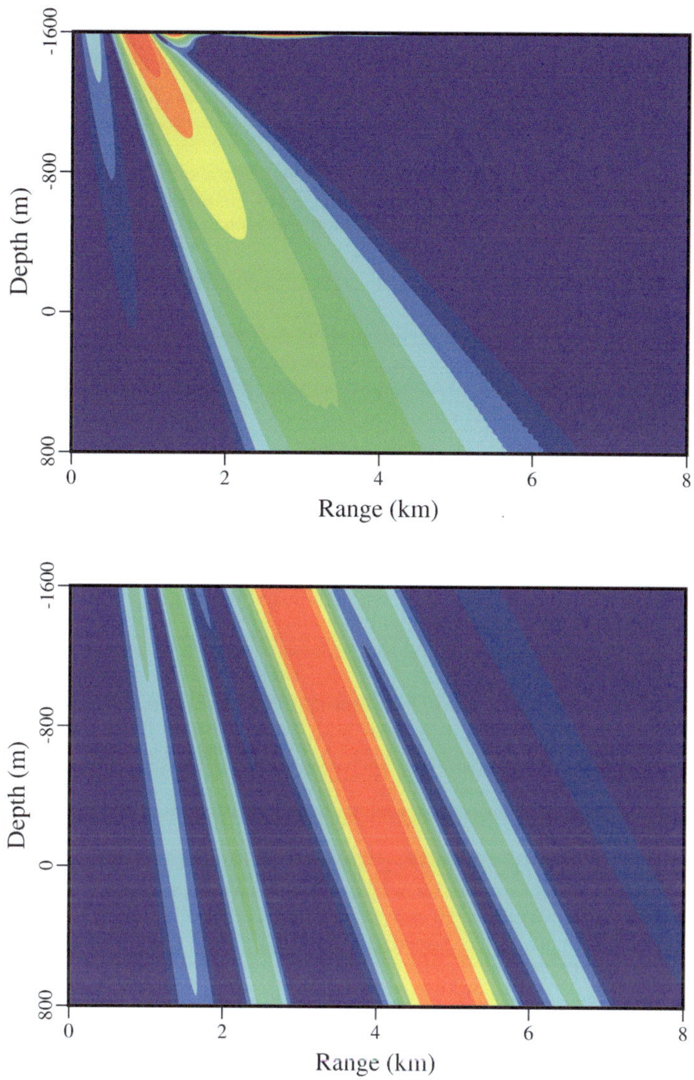

Fig. 4.15 Fast wave (top) and slow wave (bottom) transmission loss (20 dB dynamic range for each plot) for a problem in which there is an array of twenty shear sources at 25 Hz in a lossless poro-elastic half space in which $c_1 = 2400$ m/s, $c_2 = 1000$ m/s, and $c_3 = 1200$ m/s. The jth source is at a depth of $z_j = (j-1)24$ m (half wavelength spacing) and has the phase factor $\exp(-ik_3 z_j \sin 35°)$

References

1. A.E. Gill, *Ocean-Atmosphere Dynamics* (Academic, New York, 1982).
2. M.D. Collins, B.E. McDonald, W.A. Kuperman, and W.L. Siegmann, "Jovian acoustics and Comet Shoemaker-Levy 9," J. Acoust. Soc. Am. **97**, 2147–2158 (1995).
3. J.F. Lingevitch, M.D. Collins, and W.L. Siegmann, "Parabolic equations for gravity and acousto-gravity waves," J. Acoust. Soc. Am. **105**, 3049–3056 (1999).
4. M.A. Biot, "General theory of three-dimensional consolidation," J. Appl. Phys. **12**, 155–164 (1941).
5. M.A. Biot, "Theory of propagation of elastic waves in a fluid-saturated porous solid," J. Acoust. Soc. Am. **28**, 168–191 (1956).
6. M.A. Biot, "Mechanics of deformation and acoustic propagation in porous media," J. Appl. Phys. **33**, 1482–1498 (1962).
7. M.A. Biot, "Generalized theory of acoustic propagation in porous dissipative media," J. Acoust. Soc. Am. **34**, 1254–1264 (1962).
8. R. Burridge and J.B. Keller, "Poroelasticity equations derived from microstructure," J. Acoust. Soc. Am. **70**, 1140–1146 (1981).
9. T.J. Plona, "Observation of a second bulk compressional wave in a porous medium at ultrasonic frequencies," Appl. Phys. Lett. **36**, 259–261 (1980).
10. M.D. Collins, W.A. Kuperman, and W.L. Siegmann, "A parabolic equation for poro-elastic media," J. Acoust. Soc. Am. **98**, 1645–1656 (1995).
11. A.D. Pierce, "Wave equation for sound in fluids with unsteady inhomogeneous flow," J. Acoust. Soc. Am. **87**, 2292–2299 (1990).
12. A.H. Nayfeh, *Introduction to Perturbation Techniques* (Wiley-Interscience, New York, 1981), pp. 412–414.
13. G.F. Lindal, G.E. Wood, G.S. Levy, J.D. Anderson, D.N. Sweetnam, H.B. Hotz, B.J. Buckles, D.P. Holmes, P.E. Doms, V.R. Eshleman, G.L. Tyler, and T.A. Croft, "The atmosphere of Jupiter: An analysis of the Voyager Radio Occultation Measurements," J. Geophys. Res. **86**, 8721–8727 (1981).
14. S.S. Limaye, "Jupiter: New estimates of the mean zonal flow at the cloud level," Icarus **65**, 335–352 (1986).
15. J.A. Colosi, S.M. Flatté, and C. Bracher, "Internal-wave effects on 1000-km oceanic acoustic pulse propagation: Simulation and comparison with experiment," J. Acoust. Soc. Am. **96**, 452–468 (1994).
16. C. Garrett and W. Munk, "Space-time scales of internal waves: A progress report," J. Geophys. Res. **80**, 291–297 (1975).
17. *U.S. Standard Atmosphere* (U.S. Government Printing Office, Washington, 1976).
18. S.P. Timoshenko and J.N. Goodier, *Theory of Elasticity, Third Edition* (McGraw-Hill, New York, 1970), pp. 39–40.

Index

M. D. Collins, W. L. Siegmann, *Parabolic Wave Equations with Applications*,
https://doi.org/10.1007/978-1-4939-9934-7

MIX
Papier aus verantwortungsvollen Quellen
Paper from responsible sources
FSC® C105338

If you have any concerns about our products,
you can contact us on
ProductSafety@springernature.com

In case Publisher is established outside the EU,
the EU authorized representative is:
Springer Nature Customer Service Center GmbH
Europaplatz 3, 69115 Heidelberg, Germany

Printed by Libri Plureos GmbH
in Hamburg, Germany